Test Item File

GARY HOLLIS

Organic Chemistry

PAULA YURKANIS BRUICE

THIRD EDITION

Prentice Hall

Upper Saddle River, NJ 07458

Project Manager: Kristen Kaiser
Special Projects Manager: Barbara A. Murray
Production Editor: Jeanne Molenaar
Supplement Cover Manager: Paul Gourhan
Supplement Cover Designer: PM Workshop Inc.
Manufacturing Buyer: Lisa McDowell

ISBN 0-13-031012-3

Prentice-Hall International (UK) Limited, London
Prentice-Hall of Australia Pty. Limited, Sydney
Prentice-Hall Canada, Inc., Toronto
Prentice-Hall Hispanoamericana, S.A., Mexico
Prentice-Hall of India Private Limited, New Delhi
Pearson Education Asia Pte. Ltd., Singapore
Prentice-Hall of Japan, Inc., Tokyo
Editora Prentice-Hall do Brazil, Ltda., Rio de Janeiro

Contents

Section Key:

Section 98 = Questions based on the introductory material that precedes Section 1 of a chapter.

Section 99 = Cumulative questions which are (a) questions that rely on an integrated understanding of the chapter's material (ie, not found specifically in one section of the chapter) or (b) questions which rely heavily on material from a previous chapter but are pertinent to a thorough understanding of a topic in the chapter under consideration.

Prentice Hall: Instructor Support for Test Item Files

This hard copy of the test item file is just one part of Prentice Hall's comprehensive testing support service, which also includes:

• **Prentice Hall Custom Test:** This powerful computerized testing package is designed to operate on the DOS, WINDOWS, and MACINTOSH platforms. It offers full mouse support, complete question editing capabilities, random test generation, graphics, and printing capabilities.

Prentice Hall Custom Test has a unique two-track design—*Easytest*, for the novice computer user and *Fulltest*, for those who wish to write their own questions and create their own graphics.

In addition to traditional printing capabilities, Prentice Hall Custom Test offers the On-Line Testing System—the most efficient, timesaving examination aid on the market. With just a few keystrokes, the instructor can administer, correct, record, and return computerized exam results over a variety of networks.

Prentice Hall Custom Test is designed to assist educators in the recording and processing of results from student exams and assignments. Much more than a computerized gradebook, it combines a powerful database with analytical capabilities so the instructor can generate a full set of statistics. There is no grading system easier to use.

• **Prentice Hall Test Manager:** This powerful computerized testing package is designed for Windows 95, Windows 98, and Windows NT. It offers full mouse support, complete question editing capabilities, random test generation, graphics, and printing capabilities.

Prentice Hall Test Manager has removed the guesswork from your next move by incorporating "Screen Wizards" in all five databases. Each one will walk you through the important tasks from start to finish.

In addition to traditional printing capabilities, Prentice Hall Test Manager offers the On-Line Testing System—the most efficient, timesaving examination aid on the market. With just a few keystrokes, the instructor can publish a test to a local area network, correct, record and return computerized exam results over a variety of LANs.

Prentice Hall Test Manager is designed to assist educators in the recording and processing of results from student exams and assignments. Much more than a computerized gradebook, it combines a powerful database with analytical capabilities so the instructor can generate full sets of various class and individual statistics that allow you to analyze the performance of: test questions, students, an individual class or section, a course with multiple sections, and assessment types such as homework and on-line tests.

• The Prentice Hall Custom Test and Test Manager are free. To order a specific Prentice Hall Custom Test or Test Manager, you may contact your local sales representative or all our Facilities Support Services Department at 1-800-526-0485. Please identify the main text author and title.

Toll-free **technical support** is offered to all users at **1-800-550-1701**.

For those instructors without access to a computer, we offer the popular Prentice Hall Telephone Testing Service: It's simple, fast, and efficient. Simply pick the questions you'd like on your first test bank, and call the Testing Service at 1-800-550-1701; outside the U.S. and Canada, call 612-550-1705.

Identify the main text and test questions you'd like as well as any special instructions. We will create the test (or multiple versions if you wish) and send you a master copy for duplication within 48 hours. Free to adopters for the life of the text.

Chapter 1: Electronic Structure and Bonding, Acids and Bases

MULTIPLE CHOICE

1. Which of the following elements does this electronic configuration represent?

$$1s^2 \ 2s^2 \ 2p^5$$

   a. F
   b. C
   c. N
   d. Al
   e. O

   Answer: a Section: 2

2. Which carbon(s) in the following molecule is (are) sp hybridized?

$$\underset{1}{CH_3}-\underset{2}{C}-\underset{3}{CH_2}-\underset{4}{C}\equiv\underset{5}{C}-H$$

with O double bonded to carbon 2.

   a. carbon 1
   b. carbon 2
   c. carbon 1, 3
   d. carbon 4
   e. carbon 4, 5

   Answer: e Section: 9

3. Which of the following molecules <u>does not</u> exhibit a net dipole moment of zero?

   a. CO_2
   b. CH_4
   c. CCl_4
   d. H_2O
   e. SO_3

   Answer: d Section: 15

4. The conjugate acid of H_2O is:

   a.

   $H_2O:$

   b. $H_3O:$
   c. ⊕
   $H_3O:$
   d.

   $-:OH$

   e. ••⊕
   $H_2O:$

   Answer: c   Section: 16

5. Which of the following is the strongest acid?

   a. I
   b. II
   c. III
   d. IV
   e. V

   I.

   II.

   III.

   IV.

   V.

   Answer: e   Section: 18

6. Which of the following is an sp^2 hybridized carbon?

    a. $\oplus$
       CH_3
    b. $\cdot CH_3$
    c. $\ominus$
       $\colon CH_3$
    d. a and b
    e. a, b and c

Answer: d Section: 10

7. Which bond in the following molecule is the shortest?

$$CH_3 \underset{1}{-} CH_2 \underset{2}{-} CH \underset{3}{=} CH \underset{4}{-} C \underset{5}{\equiv} C - \overset{\displaystyle O}{\overset{\|}{C}} - H$$

    a. bond 1
    b. bond 2
    c. bond 3
    d. bond 4
    e. bond 5

Answer: e Section: 14

8. Which of the following molecules has a net dipole moment of zero?

    a.
```
Cl      Cl
  \    /
   C=C
  /    \
 H      H
```
    b.
```
 H      Cl
  \    /
   C=C
  /    \
Cl      H
```
    c.
```
 H      Cl
  \    /
   C=C
  /    \
 H      Cl
```
    d.
```
Cl      Cl
  \    /
   C=C
  /    \
 H      Cl
```
    e.
```
 H      H
  \    /
   C=C
  /    \
Cl      H
```

Answer: b Section: 15

9. Which of the following molecules has the smallest dipole moment?

   a. Br_2
   b. NH_3
   c. HCl
   d. HBr
   e. HI

   Answer: a   Section: 15

10. Which of the following is not a conjugate acid-base pair?

    a.
       $H_2O:$, $:OH$

    b.
       $H_2O:$, H_3O^+
    c. HSO^-_4 , H_2SO_4
    d. ^-OH, O^{2-}
    e. NO_3^- , NO_2^-

    Answer: e   Section: 16

11. What is the product formed from the following acid-base reaction?

    $$CH_3OH + :NH_3 \rightleftharpoons$$

    a. $CH_3O^- + {}^+NH_4$
    b. $CH_2OH + {}^+NH_3$
    c. $CH_3OH_2{}^+ + {}^-NH_2$
    d. $CH_3NH_2 + H_2O$
    e. $CH_4 + NH_2OH$

    Answer: a   Section: 16

12. What is the pH of a 0.1 M solution of HCl? (note: pK_a for HCl is -6)

    a. 6
    b. -6
    c. 1
    d. -8
    e. -1

    Answer: c   Section: 17

4

13. How many unpaired electrons are present in the isolated carbon atom (atomic number = 6)?

    a. none
    b. one
    c. two
    d. three
    e. four

Answer: c Section: 2

14. Which of the following is the most likely electronic structure for C_2H_2?

    a. H—C=C:
             |
             H

    b.
         ··   ··
      H—C—C:
           | ··
           H

    c.
        ··   ··
      H—C—C—H

    d. H—C≡C—H

    e.
        ⊕   ··
      H—C=C—H

Answer: d Section: 4

15. Which of the following is closest to the C-O-C bond angle in CH_3-O-CH_3?

    a. 180°
    b. 120°
    c. 109.5°
    d. 90°
    e. 160°

Answer: c Section: 11

16. What is the predicted shape, bond angle, and hybridization for $^+CH_3$?

    a. Trigonal planar, 120°, sp^2
    b. Trigonal planar, 120°, sp^3
    c. Trigonal planar, 109.5°, sp^2
    d. Trigonal pyramidal, 120°, sp^2
    e. Trigonal pyramidal, 109.5°, sp^2

Answer: a Section: 10

17. Which of the following is the strongest acid?

    a. HF
    b. H_2O
    c. $:NH_3$
    d. CH_4
    e. CH_3OH

    Answer: a   Section: 18

18. Which of the following is the strongest acid?

    a. CH_3CH_2OH
    b. CH_3OCH_3
    c. $CH_3-NH-CH_3$
    d. $CH_3-C\equiv CH$
    e. $CH_3-CH=CH_2$

    Answer: a   Section: 18

19. Which of the following ions is the strongest acid?

    a. $:H^{\ominus}$

    b. $HO:^{\ominus}$

    c. $HSO_4^{\ominus}$

    d. H_2O

    e. $H_3O^{\oplus}$

    Answer: e   Section: 16

20. Which species act as bases in the following reaction?

$$H_2SO_4 + HNO_3 \rightleftharpoons H_2NO_3^{\oplus} + HSO_4^{\ominus}$$
$$\quad 1 \qquad\; 2 \qquad\quad 3 \qquad\quad 4$$

    a. 1 and 2
    b. 3 and 4
    c. 2 and 4
    d. 1 and 3
    e. 2 and 3

    Answer: c   Section: 16

21. What is the conjugate acid of NH_3?

    a. $^+NH_3$
    b. ^-NH
    c. $^+NH_4$
    d. $^-NH_2$
    e. $^+NH_2$

    Answer: c Section: 16

22. Which of the following species have tetrahedral bond angles?

    $$\underset{A}{NH_3} \quad \underset{B}{BF_3} \quad \underset{C}{^{\oplus}CH_3} \quad \underset{D}{^{\ominus}CH_3} \quad \underset{E}{H_2O} \quad \underset{F}{H_3O^{\oplus}}$$

    a. A, D and E
    b. A, D, E and F
    c. A and E
    d. D only
    e. A, B and E

    Answer: b Section: 14

23. Which of the following is the electronic configuration of the element Fe?

    a. $1s^2\ 2s^2\ 2p^6\ 3s^2\ 3p^6\ 4s^2\ 3d^6$
    b. $1s^2\ 2s^2\ 2p^6\ 3s^2\ 3p^8\ 3d^6$
    c. $1s^2\ 2s^2\ 2p^8\ 3s^2\ 3p^6\ 4s^2\ 3d^6$
    d. $1s^2\ 2s^2\ 2p^6\ 3s^2\ 3p^6\ 4s^2\ 4d^6$
    e. $1s^2\ 2s^2\ 2p^6\ 3s^2\ 3p^6\ 4s^2\ 4p^6$

    Answer: a Section: 2

24. Which of the following structures, including formal charges, is correct for diazomethane, CH_2N_2?

    a. $:CH_2\!-\!N\!\equiv\!N:$
    b.
       $$CH_2\!=\!\overset{\ominus}{N}\!=\!\overset{\oplus}{N}:$$
    c.
       $$CH_2\!=\!\overset{\oplus}{N}\!\equiv\!\overset{\ominus}{N}:$$
    d.
       $$CH_2\!=\!\overset{\oplus}{N}\!=\!\overset{\ominus}{N}:$$
    e.
       $$CH_2\!-\!\overset{+3}{N}\!-\!\overset{-3}{N}:$$

    Answer: d Section: 4

25. What are the formal charges on nitrogen and the starred oxygen atom in the following molecule?

a. N = -1, O = 0
b. N = +1, O = -1
c. N = +1, O = +1
d. N = -1, O = -1
e. N= +1, O = 0

Answer: e Section: 4

SHORT ANSWER

26. Ar, K⁺, Cl⁻ are isoelectronic elements (elements with the same number of electrons). What orbital does the last electron occupy?

Answer: 3p orbital Section: 2

27. Using the symbol δ^+ and $\delta-$, indicate the direction of the polarity in the indicated bond.

$$CH_3 \overset{..}{\underset{..}{O}} \text{—} H$$

Answer:
$$\overset{\delta- \quad \delta+}{CH_3 \overset{..}{\underset{..}{O}} \text{—} H}$$

Section: 3

28. Give the conjugate acid and the conjugate base for HSO_4^-.

Answer:
conjugate acid: H_2SO_4

conjugate base: SO_4^{-2}
Section: 16

29. If H_2O has a pK_a value of 15.7 and HF has a pK_a value of 3.2, which is a stronger base, HO^- or F^-? Explain.

    Answer:
    HO^- is a stronger base than F^- because HF is a stronger acid than H_2O, and the stronger the acid the weaker its conjugate base.
    Section: 17

30. Draw the Kekulé structure for each of the following:

    a. CH_3CH_2OH b. CH_3CHO c. $(CH_3)_3C^+$

    Answer:

    Section: 4

31. Draw the Kekulé structure and show the direction of the dipole moment for CH_2Cl_2.

    Answer:

    Section: 15

32. BF_3 has a diplole moment of zero. Propose a structure for BF_3 that is consistent with this information.

    Answer: BF_3 is trigonal planar.

    Section: 15

33. Both sigma (σ) and pi (π) bonds can be formed by overlapping p orbitals. Describe the difference.

    Answer:
    Sigma bonds are formed from the overlap of atomic orbitals along a circular axis of symmetrical nature, i.e., head-on overlap. All single bonds are sigma bonds.

    Pi bonds are formed from the overlap of atomic orbitals along a non-symmetrical (parallel) axis, i.e., side-to-side overlap. Double and triple bonds are pi bonds.
    Section: 6

34. Explain why $:NF_3$ is a weaker base than $:NH_3$.

    Answer:
    Fluorine has an electron withdrawing effect that reduces the availability of the pair of electrons on nitrogen. Thus the basicity of $:NF_3$ is less than that of $:NH_3$.
    Section: 18

35. Explain why $AlCl_3$ is a Lewis acid.

    Answer:
    A Lewis acid is an electron pair acceptor. Aluminum in $AlCl_3$ has an empty p orbital that can accommodate the pair of electrons provided by a Lewis base.
    Section: 20

MULTIPLE CHOICE

36. The atomic number of boron is 5. The correct electronic configuration of boron is:

    a. $1s^2 2s^3$
    b. $1s^2 2p^3$
    c. $1s^2 2s^2 2p^1$
    d. $2s^2 2p^3$
    e. $1s^2 2s^2 3s^1$

    Answer: c Section: 2

SHORT ANSWER

37. Atoms with the same number of protons but different numbers of neutrons are called _____.

    Answer: isotopes Section: 1

38. Draw the shape of a 2p orbital.

    Answer:

    Section: 5

MULTIPLE CHOICE

39. How many distinct p orbitals exist in the second electron shell, where n = 2?

    a. 0
    b. 1
    c. 2
    d. 3
    e. 4

    Answer: d Section: 5

SHORT ANSWER

40. Covalent bonds may be polar or nonpolar. What property of the atoms forming a given bond determines this?

Answer: Electronegativity Section: 3

MULTIPLE CHOICE

41. The compound methylamine, CH_3NH_2, contains a C-N bond. In this bond, which of the following best describes the charge on the nitrogen atom?

   a. +1
   b. slightly positive
   c. uncharged
   d. slightly negative
   e. -1

Answer: d Section: 3

42. The formal charge on nitrogen in the compound below is _____.

```
    H  H  H
    |  |  |
 H— C— C= N—H
    |
    H
```

   a. +2
   b. +1
   c. 0
   d. -1
   e. -2

Answer: b Section: 4

SHORT ANSWER

43. The Kekulé structure of pentane is shown below. Draw the condensed structural formula which corresponds to this Lewis structure.

```
    H  H  H  H  H
    |  |  |  |  |
 H— C— C— C— C— C—H
    |  |  |  |  |
    H  H  H  H  H
```

Answer: $CH_3(CH_2)_3CH_3$ Section: 4

44. Write a completed equation for the acid-base pair shown below.

$HCO_2H + {}^-NH_2 \longrightarrow$

Answer: $HCO_2H + {}^-NH_2 \longrightarrow HCO_2{}^- + NH_3$ Section: 16

45. Consider the set of compounds, NH_3, HF, and H_2O. Rank these compounds in order of increasing acidity and discuss your rationale.

Answer:
$NH_3 < H_2O < HF$

When determining relative acidity, it is often useful to look at the relative basicity of the conjugate bases. The stronger the acid, the weaker (more stable, less reactive) the conjugate base. In this case, one would look at the relative basicity of F^-, OH^-, and $NH_2{}^-$. The relative strengths of these species can be gauged based on the electronegativity of the charged atom in each base. Since fluorine is the most electronegative, F^- is the most stable, least reactive base in the group. This means that its conjugate acid, HF, is the strongest.

Section: 18

46. Would you predict trifluoromethanesulfonic acid, CF_3SO_3H, to be a stronger or weaker acid than methanesulfonic acid, CH_3SO_3H? Explain your reasoning.

Answer:
Trifluoromethanesulfonic acid is a stronger acid. Compare the strengths of the conjugate bases and remember that the weaker the base, the stronger the conjugate acid. In the case of the trifluoro derivative, the presence of the highly electronegative fluorine atoms serves to delocalize the negative charge to a greater extent. This additional delocalization makes trifluoromethanesulfonate a weaker base.

Section: 18

MULTIPLE CHOICE

47. Which of the compounds below bond predominantly via ionic bonding?

a. KCl
b. CF_4
c. NH_3
d. both a and b
e. both b and c
f. all of the above

Answer: a Section: 3

SHORT ANSWER

48. Draw condensed structures for the four compounds with formula C_3H_9N.

   Answer:
   $CH_3CH_2CH_2NH_2$
   $CH_3CH_2NHCH_3$
   $(CH_3)_2CHNH_2$
   $(CH_3)_3N$
   Section: 4

MULTIPLE CHOICE

49. Consider the interaction of two hydrogen 1s atomic orbitals of the same phase. Which of the statements below is an incorrect description of this interaction?

   a. A sigma bonding molecular orbital is formed.
   b. The molecular orbital formed is lower in energy than a hydrogen 1s atomic orbital.
   c. The molecular orbital formed has a node between the atoms.
   d. The molecular orbital formed is cylindrically symmetric.
   e. A maximum of two electrons may occupy the molecular orbital formed.

   Answer: c   Section: 6

SHORT ANSWER

50. What kind of molecular orbital (σ, σ^*, π, or π^*) results when the two atomic orbitals shown below interact in the manner indicated?

   Answer: σ_* Section: 6

51. What kind of molecular orbital (σ, σ^*, π, or π^*) results when the two atomic orbitals shown below interact in the manner indicated?

   Answer: σ Section: 6

52. What kind of molecular orbital (σ, σ*, π, or π*) results when the two atomic orbitals shown below interact in the manner indicated?

Answer: π   Section: 6

53. What kind of molecular orbital (σ, σ*, π, or π*) results when the two atomic orbitals shown below interact in the manner indicated?

Answer: π^* Section: 6

54. What kind of molecular orbital (σ, σ*, π, or π*) results when the two atomic orbitals shown below interact in the manner indicated?

Answer: σ^* Section: 6

MULTIPLE CHOICE

55. Triethylamine [$(CH_3CH_2)_3N$] is a molecule in which the nitrogen atom is _____ hybridized and the CNC bond angle is _____.

a. sp^2, >109.5°
b. sp^2, <109.5°
c. sp^3, >109.5°
d. sp^3, <109.5°
e. sp, 109.5°

Answer: d   Section: 12

56. Choose the correct hybridization for the atom indicated in the molecule below.

$CH_3CH_2CH_2CH_3$
$\uparrow$

a. sp
b. sp^2
c. sp^3
d. none of the above

Answer: c Section: 7

57. What type of bonding is <u>most</u> important in $CH_3CH_2CH_2CH_2CH_2CH_3$?

a. ionic
b. hydrogen
c. covalent
d. polar

Answer: c Section: 3

58. What is the conjugate acid of CH_3NH_2?

a. $CH_3NH_3^+$
b. CH_3NH^-
c. NH_4^+
d. NH_2^-

Answer: a Section: 16

59. What is the conjugate base of CH_3NH_2?

a. $CH_3NH_3^+$
b. CH_3NH^-
c. NH_4^+
d. NH_2^-

Answer: b Section: 16

SHORT ANSWER

60. Write a Lewis structure for the molecule given and show all formal charges.

CH_2CO

Answer:

H
C::O:
H

Section: 4

MULTIPLE CHOICE

61. The carbon-carbon double bond in ethene is _____ and _____ than the carbon-carbon triple bond in ethyne.

a. stronger; shorter
b. stronger; longer
c. weaker; shorter
d. weaker; longer
e. stronger; more polar

Answer: d   Difficulty: 2 Section: 14

62. Which of the following contain(s) polar covalent bonds?

a. NH_3
b. Na_2O
c. H_2
d. KF
e. both a and c

Answer: a   Difficulty: 2 Section: 3

63. When a small amount of hexanoic acid [$CH_3(CH_2)_4CO_2H$, pK_a~4.8], is added to a separatory funnel which contains the organic solvent diethyl ether and water with a pH of 12.0, it is found mainly in the _____ phase as _____.

    a. ether; $CH_3(CH_2)_4CO_2^-$
    b. water; $CH_3(CH_2)_4CO_2^-$
    c. ether; $CH_3(CH_2)_4CO_2H$
    d. water; $CH_3(CH_2)_4CO_2H$
    e. none of the above

    Answer: b   Difficulty: 2   Section: 20

64. When a small amount of hexanoic acid [$CH_3(CH_2)_4CO_2H$, pKa~4.8], is added to a separatory funnel which contains the organic solvent diethyl ether and water with a pH of 2.0, it is found mainly in the _____ phase as _____.

    a. ether; $CH_3(CH_2)_4CO_2^-$
    b. water; $CH_3(CH_2)_4CO_2^-$
    c. ether; $CH_3(CH_2)_4CO_2H$
    d. water; $CH_3(CH_2)_4CO_2H$
    e. none of the above

    Answer: c   Difficulty: 2   Section: 20

65. Which of the following statements correctly describes the third electron shell that surrounds the nucleus of an atom?

    a. The third shell contains only s and p atomic orbitals.
    b. The maximum number of electrons that can occupy the third shell is 18.
    c. The total number of atomic orbitals present in the third shell is 16.
    d. The third shell can contain f orbitals.
    e. none of the above.

    Answer: b   Difficulty: 2   Section: 2

SHORT ANSWER

66. Provide the mathematical equation for the dipole moment of a bond, and identify the variables.

    Answer:
    $\mu = e \cong d$, where μ is the bond dipole moment, e is the amount of charge which is separated, and d is the distance over which the charge is separated.
    Difficulty: 2   Section: 3

67. What orbitals are used to form the covalent bonds in butane ($CH_3CH_2CH_2CH_3$)?

Answer:

The carbon-carbon σ bonds are formed by the overlap of two carbon sp^3 hybrid atomic orbitals. The carbon-hydrogen σ bonds are formed by the overlap of a carbon sp^3 hybrid atomic orbital and a hydrogen s orbital.
Difficulty: 2 Section: 7

68. Why is the C—H bond in ethene ($H_2C=CH_2$) shorter and stronger than the C—H bond in ethane(CH_3CH_3)?

Answer:
The length and strength of a C—H bond depends on the hybridization of the carbon atom. The more *s* character in the hybrid orbital used by carbon to form the bond, the shorter and stronger the bond. This is because an s orbital is closer to the nucleus than is a *p*. Ethene uses carbon sp^2 hybrid orbitals (1/3 *s* character) to make its carbon-hydrogen bonds while ethyne uses carbon *sp* (1/2 *s* character).
Difficulty: 3 Section: 14

69. At what pH will 25% of a compound with a pKa of 5.3 be in its basic form?

Answer: 4.8 Difficulty: 3 Section: 20

70. Draw the structure of a molecule which contains only carbon and hydrogen atoms (only three of which are carbon) and in which two of the carbons are sp^2 hybridized and the other is sp hybridized.

Answer: $H_2C=C=CH_2$ Difficulty: 3 Section: 14

Chapter 2: An Introduction to Organic Compounds

MULTIPLE CHOICE

1. What is the common name for the following structure?

$$CH_3 — CH — CH_3$$
$$|$$
$$CH_3$$

   a. Isobutane
   b. Isopropylmethane
   c. t-Butane
   d. n-Butane
   e. sec-Butane

   Answer: a   Section: 2

2. What is the common name for the following structure?

   a. Isobutyl bromide
   b. t-Butyl bromide
   c. Neobutyl bromide
   d. sec-Butyl bromide
   e. Isopropyl methyl bromide

   Answer: b   Section: 4

3. How many constitutional isomers are possible for C_6H_{12}?

   a. 4
   b. 5
   c. 6
   d. 7
   e. 8

   Answer: b   Section: 98

4. Give the IUPAC name for the following structure:

   a. 2-methyl-3-ethylheptane
   b. 3-ethyl-2-methylheptane
   c. 5-Isopropyloctane
   d. 4-Isopropyloctane
   e. 2-methyl-3-propylheptane

Answer: d   Section: 2

5. Give the IUPAC name for the following compound:

   a. 1-chloro-2-methylcyclohexane
   b. 1-methyl-2-chlorocyclohexane
   c. 1-chloro-5-methylcyclohexane
   d. 1-methyl-5-chlorocyclohexane
   e. 1,2-chloromethylcyclohexane

Answer: a   Section: 4

6. Which of the following is diisopropyl ether?

   a. $CH_3CH_2CH_2{-}O{-}CH_2CH_2CH_3$
   b. $CH_3CH{-}O{-}CH_2CH_2CH_3$
             $|$
           CH_3
   c. $CH_3CH_2CH_2{-}O{-}CHCH_3$
                        $|$
                      CH_3
   d.
$$CH_3 - \overset{\overset{\displaystyle CH_3}{|}}{\underset{\underset{\displaystyle CH_3}{|}}{C}} - O - \overset{\overset{\displaystyle CH_3}{|}}{\underset{\underset{\displaystyle CH_3}{|}}{C}} - CH_3$$
   e.
$$\underset{CH_3}{\overset{CH_3}{\diagup}}CH - O - \underset{\underset{\displaystyle CH_3}{|}}{CH} - CH_3$$

Answer: e   Section: 5

7. Give the IUPAC name for the following structure:

   a. 3-chloro-2-methylcyclohexanol
   b. 2-methyl-5-chlorocyclohexanol
   c. 1-chloro-4-methylcyclohexanol
   d. 5-chloro-2-methylcyclohexanol
   e. 2-methyl-3-chlorocyclohexanol

   Answer: d Section: 6

8. Which of the following is a tertiary amine?

   Answer: e Section: 1

9. Which of the following will have the lowest boiling point?

    a. CH_3Cl
    b. CH_4
    c. CH_2Cl_2
    d. $CHCl_3$
    e. CCl_4

Answer: b Section: 9

10. Which of the following has the greatest van der Waal's interaction between molecules of the same kind?

    a. $CH_3 \!-\! CH \!-\! CH_2CH_3$
               |
               CH_3
    b. $CH_3CH_2CH_2CH_3$
    c. CH_3
              |
      $CH_3 \!-\! C \!-\! CH_3$
              |
             CH_3
    d. $CH_3CH_2CH_2CH_2CH_3$
    e. $CH_3 \!-\! CH \!-\! CH \!-\! CH_3$
              | |
         CH_3 CH_3

Answer: d Section: 9

11. Which of the following has the lowest boiling point?

    a. $CH_3CH_2CH_2CH_2CH_2CH_3$
    b. $CH_3CH_2CH_2CHCH_3$
                 |
                CH_3
    c. CH_3
           |
     $CH_3CHCHCH_3$
          |
         CH_3
    d. $CH_3CH_2CHCH_2CH_3$
              |
           CH_3
    e. $CH_3CH_2CH_2CH_2CH_2CH_2CH_3$

Answer: c Section: 9

12. Which of the following has the <u>greatest</u> solubility in $CH_3CH_2CH_2CH_3$?

    a. CH_3OH
    b. $CH_3O^-Na^+$
    c. CH_3NH_2
    d. CH_3OCH_3
    e. $(CH_3)_3CH$

Answer: e Section: 9

13. Which of the following is the <u>most</u> soluble in H_2O?

    a. CH_3OCH_3
    b. CH_3CH_2OH
    c. CH_3CH_2Cl
    d. $CH_3CH_2CH_3$
    e. CH_3CHO

Answer: b Section: 9

14. Which of the following would have the <u>highest</u> boiling point?

    a. CH_3CH_2—O—CH_2CH_2–O—CH_3
    b. CH_3—O—$CH_2CH_2CH_2$–O—CH_3
    c. HO—$CH_2CH_2CH_2CH_2$—OH
    d. CH_3CH_2—O—CH_2–O–CH_2CH_3
    e. CH_3–O—CH_2CH—O—CH_3
                  |
                 CH_3

Answer: c Section: 9

15. The eclipsed and staggered forms of ethane are said to differ in:

    a. molecular formula.
    b. configuration.
    c. conformation.
    d. constitution.
    e. structure.

Answer: c Section: 10

16. Which of the following is the staggered conformation for rotation about the C_1—C_2 bond in the following structure?

   a. I
   b. II
   c. III
   d. IV
   e. V

I.

$$CH_3 \quad \overset{H}{|} \quad CH_2CH_3$$
$$H \quad \underset{H}{|} \quad H$$

$$\overset{CH_3}{\underset{1 \quad 2 \quad 3 \quad 4}{CH_3CHCH_2CH_3}}$$

II.

$$CH_3 \quad \overset{CH_3}{|} \quad CH_2CH_3$$
$$CH_3 \quad \underset{H \quad H}{}$$

III.

$$CH_3 \quad \overset{H}{|} \quad CH_3$$
$$H \quad \underset{H}{|} \quad H$$

IV.

$$C_2H_5 \quad \overset{H}{|} \quad H$$
$$H \quad \underset{H}{|} \quad H$$

V.

$$\overset{H}{\underset{H}{|}}$$
$$CH_3 \quad C_2H_5$$
$$H \qquad H$$

Answer: a   Section: 10

25

17. Which of the following is the <u>most</u> stable conformation of bromocyclohexane?

   a. I
   b. II
   c. III
   d. IV
   e. V

   I.

   II.

   III.

   IV.

   V.

   Answer: c   Section: 12

SHORT ANSWER

18. There is something wrong with the following name. Write the structure and correct the name: 2-ethylpropane.

   Answer:

   CH$_3$—CH—CH$_3$ The correct name is 2-methyblutane.
        |
        CH$_2$
        |
        CH$_3$

   Section: 2

19. Give the structure of isopentyl alcohol.

    Answer:

    $$CH_3 \diagdown$$
    $$CH—CH_2CH_2OH$$
    $$CH_3 \diagup$$

    Section: 6

20. Give the structure of tetramethylammonium chloride.

    Answer:

    $$\begin{array}{c} CH_3 \\ | \\ CH_3 — \overset{\oplus}{N} — CH_3 \ Cl^{\ominus} \\ | \\ CH_3 \end{array}$$

    Section: 7

21. Fluorine is more electronegative than chlorine yet the carbon-fluorine bond in CH_3-F is shorter than CH_3-Cl. Explain.

    Answer:
    Chlorine is a larger atom than fluorine and occupies a 3p rather than 2p orbital. The overlap of a $2sp^3$ orbital with a 3p orbital is not as good as the overlap of a $2sp^3$ orbital with a 2p orbital, causing the bond to be longer and weaker.

    Section: 8

22. CH_3CHCH_3
                 |
    Explain why CH_3 has a lower boiling point than $CH_3CH_2CH_2CH_3$.

    Answer:
    $CH_3CH_2CH_2CH_3$ has greater van der Waals forces because it has a greater contact area than isobutane. Therefore, the boiling point of $CH_3CH_2CH_2CH_3$ is higher.

    Section: 9

23. Primary and secondary amines exhibit hydrogen bonding; tertiary amines do not. Explain.

    Answer:
    The nitrogen in a tertiary amine is not attached to a hydrogen. Recall that for a molecule to exhibit hydrogen bonding it must have a hydrogen attached to a highly electronegative atom such as F, N, or O.
    Section: 9

24. Draw the Newman structure for the most stable conformation of 1-bromopropane considering rotation about the C_1-C_2 bond.

    Answer:

    Section: 10

25. Draw all possible constitutional isomers for C_2H_6O and give common names for each structure.

    Answer:

    Ethyl alcohol   Dimethyl ether
    Section: 5

26. Explain why trimethylamine, $(CH_3)_3N$:, has a considerably lower boiling point than propylamine $CH_3CH_2CH_2NH_2$, even though both compounds have the same molecular formula.

    Answer:
    Since hydrogen bonding is possible for propylamine and not for trimethylamine, the boiling point is higher for propylamine.
    Section: 9

27. Which of the molecules below has the higher boiling point? Briefly explain your choice.

$CH_3CH_2CH_2OH$ or $CH_3CH_2OCH_3$

Answer:
$CH_3CH_2CH_2OH$ has the higher boiling point since it is capable of intermolecular hydrogen bonding.

Section: 9

28. Would you expect sodium chloride (NaCl) to be highly soluble in the organic solvent hexane ($CH_3CH_2CH_2CH_2Ch_2CH_3$)? Briefly explain your answer.

Answer:
One would <u>not</u> expect NaCl to be highly soluble in hexane. NaCl is an ionic solid (i.e., a very polar material) while hexane is nonpolar. Nonpolar solvent molecules do not solvate ions well. The attractions of oppositely charged ions to each other are vastly greater than the weak attractions of the ions for the solvent.

Section: 9

29. Which compound is more soluble in water? Briefly explain your choice.

CH_3OCH_3 or CH_3CH_2OH

Answer:
CH_3CH_2OH is more soluble in water since it can donate a hydrogen bond to water and accept a hydrogen bond from water. CH_3OCH_3 can only accept a hydrogen bond from water; it has no hydrogen which can hydrogen bond to water.

Section: 9

30. Which compound is more soluble in water? Briefly explain your choice.

$(CH_3)_2NH$ or $CH_3CH_2CH_3$

Answer:
$(CH_3)_2NH$ is more soluble in water since it can hydrogen bond with water. Alkanes are not capable of hydrogen bonding with water.

Section: 9

MULTIPLE CHOICE

31. If an acyclic alkane hydrocarbon contains n carbon atoms, how many hydrogen atoms must it also contain?

    a. n
    b. n + 2
    c. n - 2
    d. 2n
    e. 2n + 2
    f. 2n - 2

    Answer: e Section: 98

SHORT ANSWER

32. Give structures for the three isomers with molecular formula C_5H_{12} and provide the common name of each.

    Answer:

    pentane or n-pentane   isopentane   neopentane

    Section: 2

33. Provide an acceptable name for the alkane shown below.

    $CH_3CH_2CH_2CH_2CH_2CH_3$

    Answer: hexane <u>or</u> n-hexane Section: 2

34. Provide an acceptable name for the alkane shown below.

    $CH_3-CH-CH_2CH_2CH(CH_3)_2$
        |
        CH_2CH_3

    Answer: 2, 5-dimethylheptane Section: 2

35. Provide an acceptable name for the alkane shown below.

Answer:
5-sec-butyl-2, 2-dimethylnonane <u>or</u>
2, 2-dimethyl-5-(1-methylpropyl) nonane
Section: 2

36. Provide an acceptable name for the alkane shown below.

$$CH_3CH_2CH_2CH_2CH_2CH_2\underset{\underset{CH(CH_3)_2}{|}}{C}HCH_2CH_2CH_3$$

Answer: 4-isopropyldecane <u>or</u> 4-(1-methlyethyl) decane   Section: 2

37. Provide an acceptable name for the alkane shown below.

$$CH_3CH_2CH_2\underset{\underset{CH_3}{|}}{C}H\underset{\underset{\underset{\underset{CH_3}{|}}{CH_2CH_2CH_3}}{|}}{C}HCH_2\underset{\underset{CH_2CH_3}{|}}{C}HCH_2CH_3$$

Answer: 3-ethyl-6-methyl-5-propylnonane   Section: 2

38. Provide an acceptable name for the alkane shown below.

$$CH_3CH_2CH_2 — \underset{\underset{CH_3CH_2CH_3}{|}}{\overset{\overset{CH_3CH_2CH_3}{|}}{C}} — \underset{\underset{CH_3CH_2CH_3}{|}}{\overset{\overset{|}{}}{C}} — H$$

Answer: 3 Ethyl-4, 4-dimethylheptane   Section: 2

39. Draw an acceptable structure for 4-t-butyloctane.

Answer:

CH$_3$ ∿∿ CH$_3$

C(CH$_3$)$_3$

Section: 2

40. Draw an acceptable structure for 3-ethyl-3-methylhexane.

Answer:

$$CH_3CH_2CH_2 - \overset{\overset{\displaystyle CH_3}{|}}{\underset{\underset{\displaystyle CH_2CH_3}{|}}{C}} - CH_2CH_3$$

Section: 2

41. Draw an acceptable structure for 4-isopropyl-2-methylheptane.

Answer:

$$CH_3\overset{\overset{\displaystyle CH_3}{|}}{C}HCH_2\underset{\underset{\displaystyle CH(CH_3)_2}{|}}{C}HCH_2CH_2CH_3$$

Section: 2

42. Draw an acceptable structure for 6-ethyl-2,6,7-trimethyl-5-propylnonane.

Answer:

CH$_3$ CH$_3$ CH$_3$ CH$_3$ CH$_3$ CH$_3$ CH$_3$ CH$_3$

Section: 2

43. Provide an acceptable name for the alkane shown below.

$$CH_3 - \underset{\underset{H}{|}}{\overset{\overset{C(CH_3)_3}{|}}{C}} - CH_2CH_2CH(CH_3)_2$$

Answer: 2,2,3,6-tetramethylheptane Section: 2

44. Provide an acceptable name for the alkane shown below.

$$CH_3CH_2CH_2CH_2 - \underset{\underset{CH_3CH_2}{|}}{\overset{\overset{H}{|}}{C}} - \underset{\underset{H}{|}}{\overset{\overset{CH_2CH_2CH(CH_3)_2}{|}}{C}} - CH_2CH_2CH_3$$

Answer: 6-ethyl-2-methyl-5-propyldecane Section: 2

45. Which intermolecular force is primarily responsible for the interactions among alkane molecules?

Answer: Van der Waal's or London forces Section: 9

MULTIPLE CHOICE

46. Consider the three isomeric alkanes n-hexane, 2,3-dimethylbutane, and 2-methylpentane. Which of the following correctly lists these compounds in order of increasing boiling point?

a. 2,3-dimethylbutane < 2-methylpentane < n-hexane
b. 2-methylpentane < n-hexane < 2,3-dimethylbutane
c. 2-methylpentane < 2,3-dimethylbutane < n-hexane
d. n-hexane < 2-methylpentane < 2, 3-dimethylbutane
e. n-hexane < 2,3-dimethylbutane < 2-methylpentane

Answer: a Section: 9

SHORT ANSWER

47. Draw a Newman projection of the most stable conformation of 2-methylpropane.

Answer:

Section: 10

48. Define the term <u>conformation</u>.

Answer:
Conformations are different arrangements of the same molecule formed by rotations about single bonds.
Section: 10

49. Use a sawhorse structure to depicit the eclipsed conformer of ethane.

Answer:

Section: 10

50. View a butane molecule along the C_2-C_3 bond and provide a Newman projection of the lowest energy conformer.

Answer:

Section: 10

51. Provide a representation of the <u>gauche</u> conformer of butane.

Answer:

Section: 10

MULTIPLE CHOICE

52. Among the butane conformers, which occur at energy minima on a graph of potential energy versus dihedral angle?

   a. gauche only
   b. eclipsed and totally eclipsed
   c. gauche and anti
   d. eclipsed only
   e. anti only

   Answer: c   Section: 10

53. Which of the following describes the most stable conformation of trans-1-tert-butyl-3-methylcyclohexane?

    a. Both groups are equatorial.
    b. Both groups are axial.
    c. The tert-butyl group is equatorial.
    d. The tert-butyl group is axial and the methyl group is equatorial.
    e. None of the above

    Answer: c Section: 14

54. Which of the following correctly ranks the cycloalkanes in order of increasing ring strain per methlene?

    a. cyclopropane < cyclobutane < cyclohexane < cycloheptane
    b. cyclohexane < cyclopentane < cyclobutane < cyclopropane
    c. cyclopentane < cyclobutane < cyclopentane < cyclopropane
    d. cyclopentane < cyclopropane < cyclobutane < cyclohexane
    e. cyclopropane < cyclopentane < cyclobutane < cyclohexane

    Answer: b Section: 11

SHORT ANSWER

55. Describe the source of angle strain and torsional strain present in cyclopropane.

    Answer:
    The angle strain arises from the compression of the ideal tetrahedral bond angle of 109.5° to 60°. The large torsional strain occurs since all C-H bonds on adjacent carbons are eclipsed.
    Section: 11

56. Draw the chair conformer of cyclohexane. Label the axial hydrogens (H_a) and the equatorial hydrogens (H_e).

    Answer:

    Section: 12

MULTIPLE CHOICE

57. Which of the following correctly lists the conformations of cyclohexane in order of increasing energy?

    a. chair < boat < twist-boat < half-chair
    b. half-chair < boat < twist-boat < chair
    c. chair < twist-boat < half-chair < boat
    d. chair < twist-boat < boat < half-chair
    e. half-chair < twist-boat < boat < chair

    Answer: d Section: 12

58. In the boat conformation of cyclohexane, the "flagpole" hydrogens are located:

    a. on the same carbon.
    b. on adjacent carbons.
    c. on C-1 and C-3.
    d. on C-1 and C-4.
    e. none of the above

    Answer: d Section: 12

59. Which of the following best explains the relative stabilities of the eclipsed and staggered forms of ethane? The _____ form has the most _____ strain.

    a. eclipsed; steric
    b. eclipsed; torsional
    c. staggered; steric
    d. staggered; torsional

    Answer: b Section: 10

60. Which of the following best explains the reason for the relative stabilities of the conformers shown?

    a. I has more torsional strain.
    b. I has more steric strain.
    c. II has more torsional strain.
    d. II has more steric strain.

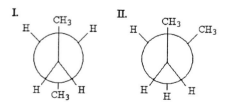

    Answer: d Section: 10

61. Which of the following compounds <u>does</u> <u>not</u> have the molecular formula $C_6H_{14}O$?

    a. 1-hexanol
    b. 2-hexanol
    c. 3-methyl-2-pentanol
    d. 3-methyl-3-pentanol
    e. cyclohexanol

    Answer: e   Difficulty: 2 Section: 6

62. What is the strongest intermolecular force present in liquid ethanol?

    a. induced dipole-induced dipole
    b. dipole-dipole, specifically hydrogen bonding
    c. dipole-dipole, but not hydrogen bonding
    d. ion-dipole
    e. ion-ion

    Answer: b   Difficulty: 2 Section: 9

63. Which of the following statements about the conformers that result from rotation about the C2-C3 bond of butane is correct?

    a. The highest energy conformer is one in which methyl groups are eclipsed by hydrogens.
    b. The gauche conformer is an eclipsed one.
    c. Steric strain is absent in the eclipsed forms.
    d. Torsional strain is absent in the eclipsed forms.
    e. none of the above.

    Answer: e   Difficulty: 2 Section: 10

SHORT ANSWER

64. The equilibrium constant for the ring-flip of fluorocyclohexane is 1.5 at 25° C. Calculate the percentage of the axial conformer at the temperature.

    Answer:
    $$K_{eq} = \frac{[eq]}{[ax]} = 1.5$$

    $$\% \text{ axial} = \frac{[ax]}{[eq] + [ax]} \cong 100\%$$

    $$= \frac{[1]}{[1.5] + [1]} \cong 100\%$$

    $$= 40\%$$

    Difficulty: 3 Section: 13

65. Draw an acceptable structure for sec-butylcyclopentane.

Answer:

Difficulty: 2 Section: 3

MULTIPLE CHOICE

66. There are 8 isomers that have the molecular formula $C_5H_{11}Br$. How many of these are tertiary alkyl bromides?

a. 0
b. 1
c. 2
d. 3
e. 8

Answer: b Difficulty: 2 Section: 1

SHORT ANSWER

67. What is polarizability and how is it related to the size of an atom?

Answer:
Polarizability indicates the ease with which an electron cloud can be distorted. The larger the atom, the more loosely its nucleus holds the electrons in its outermost shell, and the more they can be distorted.
Difficulty: 3 Section: 9

68. Draw the most stable conformer of *cis*-1-isopropyl-2-methylcyclohexane.

Answer:

Difficulty: 3 Section: 14

MULTIPLE CHOICE

69. Assuming roughly equivalent molecular weights, which of the following would have the highest boiling point?

   a. a tertiary amine
   b. a qauternary ammonium salt
   c. an alcohol
   d. an ether
   e. an alkyl chloride

   Answer: b   Difficulty: 3 Section: 15

SHORT ANSWER

70. Draw the most stable conformation of *trans*-decalin.

   Answer:

   Difficulty: 2 Section: 15

Chapter 3:   Reactions of Alkenes, Thermodynamics and Kinetics

MULTIPLE CHOICE

1. What is the common name for the following compound?

   a. t-Butylene
   b. sec-Butylene
   c. Isobutylene
   d. Butylene
   e. Methylpropylene

   Answer: c Section: 2

2. What is the IUPAC name for the following compound?

$$CH_2{=}C{-}CH_3$$
$$|$$
$$CH_2$$
$$|$$
$$CH_3$$

   a. 2-methyl-1-butene
   b. Isopentene
   c. 2-methybutene
   d. 2-Ethylpropene
   e. 3-methyl-3-butene

   Answer: a Section: 2

3. What is the IUPAC name for the following compound?

   a. 5-methylcyclohexene
   b. 4-methylcyclohexene
   c. 1-methyl-3-cyclohexene
   d. 1-methyl-4-cyclohexene
   e. methylcyclohexene

   Answer: b Section: 2

4. Which of the following is vinyl chloride?

   a. CH_3CH_2Cl
   b. $CH_2=CHCH_2Cl$
   c. $CH_2=CHCl$
   d. see graph I below
   e. $CH=CHCl$
      |
      Cl

   I. —$CH=CHCl$

   Answer: c Section: 2

5. Which of the following is capable of exhibiting cis-trans isomerism?

   a. 1-butene
   b. 1-pentene
   c. cyclohexene
   d. ethene
   e. 2-butene

   Answer: e Section: 4

6. Which of the following is <u>not</u> an electrophile?

   a. H^+
   b. BF_3
   c. $^+NO_2$
   d. Fe^{+3}
   e. $CH_2=CH_2$

   Answer: e Section: 6

7. Which of the following is <u>not</u> a nucleophile?

   a. $FeBr_3$
   b. Br^-
   c. NH_3
   d. See graph I below
   e. CH_3OCH_3

   I.

   Answer: a Section: 6

8. Which of the following statements about ethene, C_2H_4, is incorrect?

   a. The H-C-H bond angles are approximately 109.5° .
   b. All of the hydrogen atoms are in the same plane.
   c. There is a total of five sigma bonds.
   d. The carbon atoms are sp^2 hybridized.
   e. The H-C-H bond angles are approximately 120° .

   Answer: a Section: 3

9. Which of the following statements about propene, $CH_3CH{=}CH_2$, is correct?

   a. All nine atoms lie in the same plane.
   b. The compound has a cis and trans isomer.
   c. It generally acts as a Lewis acid.
   d. There is a total of eight sigma bonds.
   e. All the carbon atoms are sp^2 hybridized.

   Answer: d Section: 3

10. What is the value of ΔH in kcal/mole for the reaction shown?

    $(CH_3)_3$—C—H + Cl—Cl $\longrightarrow$ $(CH_3)_3$—C—Cl + H—Cl

    Bond energies are: $(CH_3)_3$—C—H = 91 kcal/mole
    $(CH_3)_3$—C—Cl = 78.5 kcal/mole
    Cl—Cl = 58 kcal/mole
    H—Cl = 103 kcal/mole

    a. +32.5
    b. -57.5
    c. -32.5
    d. +57.5
    e. -8.5

    Answer: c Section: 7

11. How many transition states are present in the following reaction diagram?

    a. 3
    b. 4
    c. 5
    d. 2
    e. 1

Reaction Coordinate ⟶

Answer: d Section: 7

12. An <u>increase</u> in which of the following results in a <u>decrease</u> in the rate of the chemical reaction?

    a. Temperature
    b. Concentration
    c. Collision frequency
    d. Energy of activation
    e. Fraction of collisions with proper orientation

Answer: d Section: 7

13. An increase in which of the following will occur if the reaction temperature is <u>increased</u>?

    I.   Energy of activation
    II. Collision frequency
    III. Fraction of collisions with sufficient energy

    a. I and II
    b. I and III
    c. II and III
    d. I, II, and III
    e. I

Answer: c Section: 7

14. What is the activation energy for the reaction B → A in the following diagram?

    a. A
    b. B
    c. C
    d. D
    e. E

Reaction Coordinate ⟶

Answer: e   Section: 7

15. Which of the following reagents gives the reaction shown below?

$$CH_3CH = CH_2 + ? \longrightarrow CH_3CH_2CH_3$$

    a. H_2/HCl
    b. H_2/H_2SO_4
    c. H_2/Ni
    d. H_2O/Ni
    e. H_2O/H_2SO_4

Answer: c   Section: 19

16. Which of the following is the most stable alkene?

    a. $CH_3-CHCH=CH_2$
       |
       CH_3
    b. $CH_3CH_2C=CH_2$
       |
       CH_3
    c. $CH_3C=CH_2$
       |
       CH_3
    d. H CH_3
       \   /
         C=C
       /   \
      CH_3 CH_3
    e. CH_3CH_2 CH_3
       \   /
         C=C
       /   \
      CH_3 CH_3

Answer: e   Section: 19

17. Which of the following alkenes reacts with HCl at the slowest rate?

a. $CH_3CHCH=CH_2$
    |
    CH_3

b. $CH_3CH_2C=CH_2$
    |
    CH_3

c. $CH_3C=CH_2$
    |
    CH_3

d. H⟍   ⟋CH_3
        C=C
    ⟋   ⟍
 CH_3 CH_3

e. CH_3CH_2⟍   ⟋CH_3
           C=C
        ⟋   ⟍
    CH_3 CH_3

Answer: a Section: 10

18. Which of the following is the most stable carbocation?

   a. I
   b. II
   c. III
   d. IV
   e. V

I.

II.

III.

IV.

V.

Answer: d Section: 10

19. Which of the following is a step in the mechanism of the reaction shown?

$$CH_2=CHCH_3 + HBr \xrightarrow{\text{Peroxide}} CH_2CH_2CH_3$$
$$\qquad\qquad\qquad\qquad\qquad\qquad\qquad | $$
$$\qquad\qquad\qquad\qquad\qquad\qquad\qquad Br$$

a.
   $CH_2\overset{\bullet}{C}HCH_2 + HBr$
   |
   Br

b. $\overset{\bullet}{}$
   $CH_2CH_2CH_3 + HBr$

c. $\overset{\oplus}{}$
   $CH_2CHCH_3 + HBr$
   |
   Br

d. $CH_2CH_2CH_3 + HBr$
   $\overset{\oplus}{}$

e.
   $CH_2\overset{\bullet}{C}HCH_3 + H\cdot$
   |
   Br

Answer: a Section: 18

20. Which of the following carbocations is likely to rearrange?

   a. I
   b. II
   c. III
   d. IV
   e. II and IV

I. CH_3 CH_3

II. CH_3 CH_3

III. CH_3 CH_3

IV. CH_3 $^{\oplus}CH_2$

Answer: e Section: 14

21. What is the major product of the following reaction?

    a. I
    b. II
    c. III
    d. IV
    e. V

I.

II.

III.

IV.

V.

Answer: c Section: 13

22. What is the major product of the following reaction?

    a. I
    b. II
    c. III
    d. IV
    e. V

Answer: b   Section: 15

23. What reagents are needed to accomplish the following transformation?

    a. H_2O/H^+
    b. H_2O/Peroxide
    c. $OH-$
    d. BH_3
    e. 1. BH_3 / 2. HO^-, H_2O_2, H_2O

Answer: e   Section: 17

24. Which reaction intermediate is formed when Br_2/CCl_4 reacts with cyclohexene?

    a. I
    b. II
    c. III
    d. IV
    e. V

    I.

    II.

    III.

    IV.

    V.

    Answer: d Section: 15

SHORT ANSWER

25. Under what conditions is $\Delta G°$ equal to $\Delta H°$ for a chemical reaction?

    Answer: Since $\Delta G° = \Delta H° - T\Delta S°$, then $\Delta G°$ is equal to $\Delta H°$ when $T\Delta S°$ is zero.

    Section: 7

26. Draw all the possible constitutional isomers of C_4H_8.

    Answer:

    $CH_3CH_2CH{=}CH_2$ $CH_3CH{=}CHCH_3$ $CH_3\overset{\overset{\displaystyle CH_3}{|}}{C}{=}CH_2$
    (cis trans)

    Section: 4

27. Complete the following tree of reactions by giving the major products:

Answer: a. CH₃CHCH₃
$$CH_3\overset{\displaystyle CH_3}{\underset{}{C}}HCH_3$$
         |
         CH₃

b.   CH₃
          |
      CH₃CCH₃
          |
          Cl

c.   Br
          |
      CH₃CHCH₂
          |
          CH₃

d.   CH₃
          |
      CH₃CCH₃
          |
          OH

e.   CH₃
          |
  CH₃—C—CH₂
        | |
        Br Br

f.   CH₃
        |
    CH₃CHCH₂
          |
          OH

g.   CH₃
          |
   CH₃—C—CH₂
        | |
        OH Br

53

Answer:

h.   CH$_3$
        |
    CH$_3$CCH$_3$
        |
       OH

Section: 9

28. Which of the following alkenes would have the smallest heat of hydrogenation? Explain.

$$CH_2=CH_2 \text{ or } CH_3CH=CH_2$$

Answer:
CH$_3$CH=CH$_2$ would have the smallest heat of hydrogenation because the double bond is more stable than that of CH$_2$=CH$_2$. The greater stability is due to the fact that the double bond in propane is more alkyl substituted than that of ethene.

Section: 19

29. Based on the following energy diagram, which compound, A or C, is formed faster from B? Which is more stable, A or C? Explain.

Answer:
A is formed faster since the pathway for its formation has the smaller activation energy. C is more stable than A because it has a lower energy.

Section: 7

30. Assign the E or Z configuration to the following molecule:

Answer:
The molecule has a Z configuration based on the priority rule
established by Cahn-Ingold-Prelog.

Section: 5

31. Propose a mechanism for the following reaction:

Answer:

I.

II.

III.

IV.

Section: 13

55

32. Draw the major organic product generated in the reaction below.

Answer:

Section: 12

33. Draw the major organic product generated in the reaction below.

Answer:

Section: 12

34. Draw the major organic product generated in the reaction below.

Answer:

Section: 12

35. Draw the major organic product generated in the reaction below.

Answer:

Section: 12

36. Draw the major organic product generated in the reaction below.

Answer:

Section: 12

37. Draw the major organic product generated in the reaction below.

Answer:

Section: 12

57

38. Draw the major organic product generated in the reaction below.

$$H^+, H_2O$$
dilute aqueous acid

Answer:

Section: 13

39. Draw the major organic product generated in the reaction below.

$$H^+, H_2O$$
dilute aqueous acid

Answer:

Section: 13

40. Draw the major organic product generated in the reaction below.

$$H^+, H_2O$$
dilute aqueous acid

Answer:

Section: 13

41. Draw the major organic product generated in the reaction below.

$$\text{1. Hg (OAc)}_2, \text{H}_2\text{O}$$
$$\text{2. NaBH}_4$$

Answer:

Section: 16

42. Draw the major organic product generated in the reaction below.

$$\text{1. Hg (OAc)}_2, \text{H}_2\text{O}$$
$$\text{2. NaBH}_4$$

Answer:

Section: 16

43. Draw the major organic product generated in the reaction below.

$$\text{1. BH}_3$$
$$\text{2. H}_2\text{O}_2, ^-\text{OH}$$

Answer:

Section: 17

44. Draw the major organic product generated in the reaction below.

$$\xrightarrow[\text{2. } H_2O_2, {}^-OH]{\text{1. } BH_3}$$

Answer:

OH

Section: 17

45. Draw the major organic product generated in the reaction below.

$$\xrightarrow{Br_2, CCl_4}$$

Answer:

Br

Br

Section: 15

46. Draw the major organic product generated in the reaction below.

$$\xrightarrow{Cl_2, H_2O}$$

Answer:

OH

Cl

Section: 15

47. Complete the following reaction and provide a detailed, step-by-step mechanism for the process.

CH₃ / Br₂ →

Answer:

Section: 15

48. Complete the following reaction and provide a detailed, step-by-step mechanism for the process.

HBr →

Answer:

Section: 9

49. Complete the following reaction and provide a detailed, step-by-step mechanism for the process.

H^+, H_2O →

Answer:

Section: 13

50. Provide a detailed, step-by-step mechanism for the reaction shown below.

Answer:

Section: 15

51. Based on the relative stabilities of the intermediates involved, explain the basis for Markovnikov's rule in the addition of hydrogen halides to alkenes.

Answer:
The rate-determining step in this reaction is the production of a carbocation intermediate. Since this step is endothermic, Hammond's postulate allows one to gauge the relative stabilities of the transition states by comparing the relative stabilities of the carbocation intermediates. The reaction pathway which produces the more substituted carbocation will thus occur more rapidly.

Section: 11

52. Explain the regioselectivity observed in the radical addition of HBr to 2-methylpropene.

Answer:
The reaction proceeds via the addition of Br• to the alkene. Two competing pathways are possible, but the transition state leading to the more substituted alkyl radical is lower in energy. This process ultimately makes the addition anti-Markovnikov in nature.

Section: 18

MULTIPLE CHOICE

53. Which of the following is the best reaction sequence to use if one wants to accomplish a Markovnikov addition of water to an alkene with minimal skeletal rearrangement?

    a. water + dilute acid
    b. water + concentrated acid
    c. oxymercuration-demercuration
    d. hydroboration-oxidation
    e. none of the above

    Answer: c Section: 16

SHORT ANSWER

54. Given an activation energy of 15 kcal/mol, use the Arrhenius equation to estimate how much faster the reaction will occur if the temperature is increased from 100° C to 120° C. R = 1.987 cal/mol·K.

    Answer: The reaction will occur about 2.8 times faster. Section: 7

55. Consider the conversion of C to D via a one-step mechanism. The activation energy of this conversion is 3 kcal/mol. The energy difference between D and the transition state of the reaction is 7 kcal/mol. Estimate $\Delta H°$ for the reaction C ----> D.

    Answer: 4 kcal/mol Section: 7

56. Consider the one-step conversion of F to G. Given that the reaction is endergonic by 5 kcal/mol and that the energy difference between G and the transition state for the process is 15 kcal/mol, sketch a reaction-energy profile for this reaction. Make sure to show how the given energy differences are consistent with your sketch.

    Answer:

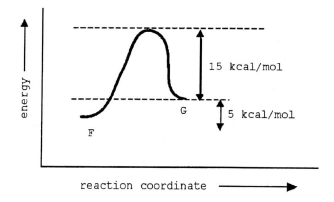

    Section: 7

57. Provide the proper IUPAC name for the alkene shown below.

$$CH_3 \quad CH_2CH_2CH_3$$
$$\diagdown \quad \diagup$$
$$C = C$$
$$\diagup \quad \diagdown$$
$$CH_3CH_2 \quad CH_3$$

Answer: (E)-3,4-dimethyl-3-heptene   Section: 2

58. Provide the proper IUPAC name for the alkene shown below.

Answer: 6-bromo-1-methylcyclohexene   Section: 2

59. Draw and name the six alkenes which have the molecular formula C_5H_{10}.

Answer:

1-pentene	trans-2-pentene or (E)-2-pentene	cis-2-pentene or (Z)-2-pentene
3-methyl-1-butene	2-methyl-1-butene	2-methyl-2-butene

Section: 2

MULTIPLE CHOICE

60. Which of the following best describes the geometry about the carbon-carbon double bond in the alkene below?

CH$_3$ Cl
  \   /
   C=C
  /   \
CH$_3$ Cl

a. E
b. Z
c. neither E nor Z

Answer: c Section: 5

SHORT ANSWER

61. The $\Delta G°$ for the conversion of "axial" isopropylcyclohexane to "equatotial" isopropylcyclohexane is -2.1 kcal/mol. Calculate the percentage of isopropylcyclohexane molecules that have the isopropyl substituent in the axial position at this temperature. [$R = 1.987 \cong 10^{-3}$ kcal mol^{-1}K^{-1}]

Answer:
$\Delta G° = -RTlnK$

-2.1kcal/mol = -(1.987 $\cong$ 10^{-3} kcal mol^{-1} K^{-1})(298K)lnK

$lnK = \dfrac{2.1 \ \text{kcal mol}^{-1}}{(1.987 \times 10^{-3} \text{ kcal mol}^{-1} \text{ K}^{-1})(298K)}$

$K = \dfrac{[eq]}{[ax]} = 34.7$

% axial = $\dfrac{[ax]}{[eq] + [ax]}$ $\cong$ 100%

= $\dfrac{[1]}{[34.7] + [1]}$ $\cong$ 100%

= 2.8%

Difficulty: 2 Section: 7

MULTIPLE CHOICE

62. Which of the following contributes to make $\Delta G°$ more negative?

    a. use of a catalyst
    b. a more positive $\Delta H°$
    c. a more positive $\Delta S°$
    d. a larger rate constant
    e. none of the above

    Answer: c   Difficulty: 2 Section: 7

SHORT ANSWER

63. What is the free energy of activation of a one-step reaction? How is it *qualitatively* related to the rate constant of the reaction?

    Answer:
    The free energy of activation of a one-step reaction is the difference in free energy between the transition state and the reactants. As the free energy of activation increases, the rate constant decreases.

    Difficulty: 2 Section: 7

MULTIPLE CHOICE

64. Which of the following correctly describes intermediates and/or transition states?

    a. Transition states occur at minima on reaction coordinate diagrams.
    b. Both transition states and intermediates occur at maxima on reaction coordinate diagrams.
    c. An intermediate is always produced after the rate-determining step of a reaction mechanism.
    d. Transition states have partially formed bonds whereas intermediates have fully formed bonds.
    e. none of the above.

    Answer: d   Difficulty: 2 Section: 7

SHORT ANSWER

65. What is hyperconjugation, and how does it affect carbocation stability?

    Answer:
    Delocalization of electrons by the overlap of a o bond with an empty p orbital is hyperconjugation. This effect occurs only when the interacting orbitals are properly aligned. The more alkyl groups which are substituents of a cationic carbon, the more the cation can be stabilized by the delocalizing hyperconjugation effects.

    Difficulty: 3 Section: 10

MULTIPLE CHOICE

66. According to the Hammond Postulate, which of the following is correct?

   a. The transition state of an endothermic reaction step will be more reactant-like than product-like.
   b. The intermediate of an endothermic reaction step will be more reactant-like than product-like.
   c. The transition state of an exothermic reaction step will be more reactant-like than product-like.
   d. All transition states are more product-like than reactant-like.
   e. None of the above.

   Answer: c   Difficulty: 2 Section: 11

67. Which of the following compounds will react most rapidly with HCl?

   a. 5-methyl-1-hexene
   b. 4-methyl-1-hexene
   c. (E)-5-methyl-2-hexene
   d. (E)-2-methyl-3-hexene
   e. 2-methyl-2-hexene

   Answer: e   Difficulty: 2 Section: 11

68. What is the major product from the acid-catalyzed hydration of 2-methyl-2-pentene?

   a. 2-methylpentane
   b. 2-methyl-1-pentanol
   c. 2-methyl-2-pentanol
   d. 2-methyl-3-pentanol
   e. 1-methoxypentane

   Answer: c   Difficulty: 2 Section: 13

SHORT ANSWER

69. Draw the structure of the major product of the reaction below.

   Answer:

   Difficulty: 3 Section: 15

70. Show the propagating steps in the addition of HBr to 1-pentene in the presence of peroxide.

Answer:

Difficulty: 3 Section: 18

Chapter 4:   Stereochemistry

MULTIPLE CHOICE

1. What type of isomers are the staggered and eclipsed forms of ethane?

   a. Constitutional
   b. Structural
   c. Configurational
   d. Conformational
   e. Stereochemical

   Answer: d Section: 1

2. What is the relationship between the following compounds?

   a. Configurational isomers
   b. Conformational isomers
   c. Constitutional isomers
   d. Structural isomers
   e. Positional isomers

   Answer: a Section: 2

3. Which of the following compounds has a chirality center?

    a. I
    b. II
    c. III
    d. IV
    e. V

I.
$$CH_3 - \overset{\displaystyle Cl}{\underset{\displaystyle H}{|}} - CH_3$$

II.
$$CH_3 - \overset{\displaystyle Cl}{\underset{\displaystyle C_2H_5}{|}} - Cl$$

III.
$$CH_3 - \overset{\displaystyle Cl}{\underset{\displaystyle Br}{|}} - H$$

IV.
$$Cl - \overset{\displaystyle Cl}{\underset{\displaystyle CH_3}{|}} - H$$

V.

Answer: c   Section: 3

4. Which of the following compounds is never chiral?

    a. 2,3-dibromobutane
    b. 1,3-dibromobutane
    c. 1,2-dichlorobutane
    d. 1,4-dibromobutane
    e. 1-bromo-2-chlorobutane

Answer: d   Section: 8

5. A and B are stereoisomers. They are nonsuperimposable and are mirror images of one another. Which of the following best describes the relationship between A and B?

    a. Structural isomers
    b. Enantiomers
    c. Confrontational isomers
    d. Diastereomers
    e. Constitutional isomers

Answer: b   Section: 3

6. What is the relationship between the following compounds?

    a. Superimposable without bond rotation
    b. Constitutional isomers
    c. Conformational isomers
    d. Diastereomers
    e. Enantiomers

Answer: e Section: 8

7. What is the relationship between the following compounds?

    a. Constitutional isomers
    b. Enantiomers
    c. Diastereomers
    d. Conformational isomers
    e. Superimposable without bond rotation

$$
\begin{array}{cc}
\overset{CH_3}{\underset{CH_3}{H{\Rightarrow}C{\Leftarrow}Cl}} & \text{and} \qquad \overset{CH_3}{\underset{H \quad CH_2Cl}{C^{\cdots\prime\prime}H}}
\end{array}
$$

Answer: a Section: 98

8. Which of the following compounds has an S configuration?

    a. Cl
         $H{-}|{-}CH_3$
         CH_2Cl
    b. CH_2Cl
         $H{-}|{-}CH_3$
         Cl
    c. CH_2Cl
         $CH_3{-}|{-}H$
             Cl
    d. H
         $CH_3{-}|{-}Cl$
             CH_2Cl
    e. CH_2Cl
         $Cl{-}|{-}CH_3$
             H

Answer: b Section: 5

9. Which of the following compounds has an R configuration?

    a.

$$CH_3CH_2 \overset{\displaystyle H}{\underset{\displaystyle CH=CH_2}{-\!\!|\!\!-Br}}$$

    b.

$$Br \overset{\displaystyle CH=CH_2}{\underset{\displaystyle CH_2CH_3}{-\!\!|\!\!-H}}$$

    c.

$$Br \overset{\displaystyle CH_2CH_3}{\underset{\displaystyle H}{-\!\!|\!\!- CH=CH_2}}$$

    d.

$$CH_2=CH \overset{\displaystyle CH_2CH_3}{\underset{\displaystyle Br}{-\!\!|\!\!-H}}$$

    e.

$$CH_3CH_2 \overset{\displaystyle CH=CH_2}{\underset{\displaystyle H}{-\!\!|\!\!-Br}}$$

Answer: a   Section: 5

10. Which of the following groups has the highest priority using the Cahn, Ingold, Prelog rules?

    a.

$$\overset{\displaystyle O}{\overset{\|}{-C}}-OH$$

    b.

$$\overset{\displaystyle O}{\overset{\|}{-C}}-H$$

    c. —OH

    d. —O—CH$_3$

    e.

$$\overset{\displaystyle O}{\overset{\|}{-C}}-O-CH_3$$

Answer: d   Section: 5

11. What is the relationship between the structures shown below?

    a. Structural isomers
    b. Constitutional isomers
    c. Conformational isomers
    d. Configurational isomers
    e. Enantiomers

Answer: c Section: 98

12. What is the relationship between the structures shown below?

    a. Diastereomers
    b. Constitutional isomers
    c. Conformational isomers
    d. Configurational isomers
    e. Enantiomers

Answer: b Section: 98

13. What is the relationship between the structures shown below?

    a. Enantiomers
    b. Diastereomers
    c. Configurational isomers
    d. Identical compounds
    e. Constitutional isomers

Answer: d Section: 3

14. Which of the following compounds is an enantiomer of the structure below?

a.   CH$_3$
     H—|— Cl
       H

b.   Cl
     H—|—H
       CH$_3$

c.   H
     Cl—|— CH$_3$
       H

d. a and b

e. It does not have an enantiomer.

       CH$_3$
  H,,,.C
  Cl   H

Answer: e Section: 3

15. Which of the following is <u>not</u> true of enantiomers?

a. They have the same melting point.
b. They have the same boiling point.
c. They have the same chemical reactivity with non-chiral reagents.
d. They have the same density.
e. They have the same specific rotation.

Answer: e Section: 3

16. The specific rotation of a pure substance is 1.68° . What is the specific rotation of a mixture containing 75% of this isomer and 25% of the (-) isomer?

a. +1.68°
b. 0°
c. +1.26°
d. +0.84°
e. +.042°

Answer: d Section: 7

17. The specific rotation of a pure substance is -5.90°. What is the percentage of this isomer in a mixture with an observed specific rotation of -2.95° ?

    a. 25%
    b. 50%
    c. 75%
    d. 80%
    e. 0%

    Answer: c Section: 7

18. What is the relationship between the structures shown below?

    a. Enantiomers
    b. Diastereomers
    c. Configurational isomers
    d. Identical compounds
    e. Constitutional isomers

    Answer: b Section: 8

19. Which of the following compounds is chiral?

    a.
      H
      CH₃—|—Cl
      CH₃—|—Cl
      H

    b.
      H
      CH₃—|—Br
      CH₃—|—Cl
      H

    c.
      CH₃
      CH₃—|—Br
      CH₃—|—Cl
      CH₃

    d.
      CH₃
      CH₃—|—Cl
      CH₃—|—Cl
      CH₃

e. CH_3
    Cl—|—Cl
    Br—|—Br
        CH_3

Answer: b Section: 3

20. Which of the following is a meso compound?

a. CH_3
    H—|—Cl
    H—|—Cl
        CH_3

b. CH_3
    H—|—Cl
    Cl—|—H
        CH_3

c. CH_3
    H—|—Cl
    H—|—Cl
        CH_2CH_3

d. CH_3
    H—|—Cl
    Cl—|—H
        CH_2CH_3

e. CH_3
    Br—|—Cl
    Br—|—Cl
        CH_2Cl

Answer: a Section: 9

21. What is the configuration of the following compound?

$$CH_3$$
H—|— Cl
Cl—|— H
$$CH_2CH_3$$

a. 2S, 3R
b. 3R, 3S
c. 2S, 3S
d. 2R, 3R
e. Can't do R and S; the compound is achiral.

Answer: c Section: 10

22. What is the relationship between the following compounds?

a. Enantiomers
b. Diastereomers
c. Constitutional isomers
d. Conformational isomers
e. Identical compounds

Answer: e Section: 9

23. What is the relationship between the following compounds?

$$CH_3$$
Cl—|— H
H—|— Cl
$$CH_3$$

and

$$CH_3$$
H—|— Cl
Cl—|— H
$$CH_3$$

a. Enantiomers
b. Diastereomers
c. Constitutional isomers
d. Conformational isomers
e. Identical compounds

Answer: a Section: 8

24. What is the relationship between the following compounds?

$$
\begin{array}{c}
CH_3 \\
Cl-|-H \\
Cl-|-H \\
CH_3
\end{array}
\qquad and \qquad
\begin{array}{c}
CH_3 \\
H-|-Cl \\
Cl-|-H \\
CH_3
\end{array}
$$

    a. Enantiomers
    b. Diastereomers
    c. Constitutional isomers
    d. Conformational isomers
    e. Identical compounds

Answer: b Section: 8

25. A sample of (R)-2-chlorobutane,

$$
\begin{array}{c}
H \\
CH_3-|-Cl \\
CH_2CH_3
\end{array}
$$

reacts with Br_2 in the presence of light, and all the products having the formula C_4H_8BrCl were isolated. Two possible isomers are shown below:

$$
\begin{array}{c}
CH_3 \\
Br-|-Cl \\
CH_2CH_3
\end{array}
\qquad and \qquad
\begin{array}{c}
CH_3 \\
Cl-|-Br \\
CH_2CH_3
\end{array}
$$

    I   II

Of these:

    a. only I was formed
    b. both I and II were formed in equal amounts
    c. both I and II were formed in unequal amounts
    d. only II was formed
    e. neither I nor II was formed

Answer: b Section: 11

26. Which of the following is/are optically inactive?

    a. A 50-50 mixture of R and S enantiomers
    b. A meso compound
    c. Every achiral compound
    d. A racemic mixture
    e. All the above

    Answer: e Section: 6

27. The configuration of R-(+)-gluceraldehyde is as follows:

$$
\begin{array}{c}
O \\
\parallel \\
C-H \\
H-\!\!\mid\!\!-OH \\
CH_2OH
\end{array}
$$

    What is the absolute configuration of (-)-lactic acid?

$$
\begin{array}{c}
O \\
\parallel \\
C-OH \\
H-\!\!\mid\!\!-OH \\
CH_2OH
\end{array}
$$

    a. R configuration
    b. L configuration
    c. S configuration
    d. R and S configuration
    e. D and L configuration

    Answer: a Section: 5

28. Which of the following compounds contain hydrogens labeled H_a and H_b that are enantiotopic?

    a. I
    b. II
    c. III
    d. IV
    e. V

I.

$$CH_3-\overset{\overset{\displaystyle H_a}{|}}{\underset{\underset{\displaystyle H_b}{|}}{C}}-CH_3$$

II.

III.

IV.

V.

Answer: d Section: 14

29. Which of the following is a meso compound?

    a. trans-1,4-dimethylcyclohexane
    b. cis-1,3-dimethylcyclohexane
    c. trans-1,3-dimethylcyclohexane
    d. cis-1,4-dimethylcyclohexane
    e. trans-1,2-dimethylcyclohexane

Answer: b Section: 9

SHORT ANSWER

30. Indicate whether each of the following structures has the R or S configuration. What is the relationship between the two structures?

$$\text{Br}$$
$$\text{HO}-\overset{\displaystyle|}{\underset{\displaystyle|}{}}-\text{CH}_3 \quad \text{and} \quad \text{CH}_3\text{CH}_2-\overset{\displaystyle|}{\underset{\displaystyle|}{}}-\text{Br}$$
$$\text{CH}_2\text{CH}_3 \qquad\qquad \text{CH}_3$$

assign priorities to each group.

Answer:

$$\text{Br}^1$$
$$\text{HO}^2-\overset{\displaystyle|}{\underset{\displaystyle|}{}}-\text{CH}_3{}^4 \qquad \text{Configuration is R}$$
$$\text{CH}_2\text{CH}_3$$

$$\text{OH}$$
$$\text{CH}_3\text{CH}_2-\overset{\displaystyle|}{\underset{\displaystyle|}{}}-\text{Br} \qquad \text{Configruation is S}$$
$$\text{CH}_3$$

Therefore, the two compounds are enantiomers.
Section: 3

31. A solution containing 0.96 g of 2-bromooctane in 10ml ether solution gave an observed rotation of -1.8° in a 10cm cell at 20° C. Calculate the specific rotation of this solution.

Answer:

$$\alpha]_D^T = \frac{\alpha}{1c}$$

$$[\alpha]_D^{20°c} = \frac{1.8°}{1\text{dm} \cong \dfrac{0.96\text{g}}{10\text{ml}}} = -18.75° \text{ in ether}$$

Section: 6

32. Consider the molecule $C_2H_2Br_2Cl_2$.

   a. Draw a structure that is optically inactive because it does not have a chirality center.
   b. Draw a structure that is optically inactive because it is a meso compound.
   c. Draw a structure that is optically active because it is chiral.

Answer:

a.
```
        Cl
        |
   Cl——|——H
   Br——|——H      no chirality center
        |
        Br
```

b.
```
        Cl
        |
    H——|——Br
    H——|——Br     meso compound
        |
        Cl
```

c.
```
        Br
        |
   Cl——|——H
    H——|——Cl     chiral
        |
        Br
```

Section: 6

33. Is the molecule shown below chiral or achiral?

Answer: achiral Section: 3

34. Is the molecule shown below chiral or achiral?

Answer: chiral Section: 8

35. Is the molecule shown below chiral or achiral?

Answer: achiral   Section: 8

36. is the molecule shown below chiral or achiral?

CH_3 ◢⟩····⊪ OH

Answer: achiral   Section: 3

37. Which of the following terms best describes the pair of compounds shown: enantiomers, diastereomers, or the same compound?

Answer: the same compound   Section: 3

38. Which of the following terms best describes the pair of compounds shown: enantiomers, diastereomers, or the same compound?

Answer: enantiomers   Section: 8

39. Which of the following terms best describes the pair of compounds shown: enantiomers, diastereomers, or the same compound?

Answer: disatereomers   Section: 8

40. Which of the following terms best describes the pair of compounds shown: enantiomers, diastereomers, or the same compound?

Answer: diastereomers Section: 8

41. Label each chiral carbon in the compound below as R or S.

Answer:

Section: 5

42. Draw the structure of (2R, 3S)-dichloropentane. Take particular care to indicate three-dimensional sterochemical detail properly.

Answer:

Section: 8

43. Draw the structure of (2S - 3R)-dichloropentane. Take particular care to indicate three-dimensional sterochemical detail properly.

Answer:

Section: 8

84

44. Draw the structure of any diastereomer of (2R, 3S)-dichloropentane. Take particular care to indicate three-dimensional sterochemical detail properly.

Answer:

or

Section: 8

45. Draw the structure of a meso form of 1,3-dichlorocyclopentane. Take particular care to indicate three-dimensional sterochemical detail properly.

Answer:

Section: 9

MULTIPLE CHOICE

46. Which of the statements below correctly describes an achiral molecule?

   a. The molecule has a nonsuperimposable mirror image.
   b. The molecule exhibits optical activity when it interacts with plane-polarized light.
   c. The molecule has an enantiomer.
   d. The molecule might be a meso form.
   e. None of the above.

Answer: d Section: 9

47. Which of the following statements correctly pertains to a pair of enantiomers?

    a. They rotate the plane of polarized light by exactly the same amount and in opposite directions.
    b. They rotate the plane of polarized light by differing amounts and in opposite directions.
    c. They rotate the plane of polarized light by differing amounts and in the same direction.
    d. They have different melting points.
    e. They have the same melting point, but they have different boiling points.

    Answer: a Section: 6

SHORT ANSWER

48. Can one predict whether a compound with a single chiral carbon is dextro- or levorotatory based on the R/S assignment at this chiral carbon? Explain briefly.

    Answer:
    No. R/S assignment is purely a convention of nomenclature and is completely independent of the direction in which plane-polairzed light is rotated by the compound.
    Section: 12

MULTIPLE CHOICE

49. If (S)-glyceraldehyde has a specific rotation of -8.7°, what is the specific rotation of (R)-glyceraldehyde?

    a. -8.7°
    b. +8.7°
    c. 0.0°
    d. cannot be determined from the information given

    Answer: b Section: 6

SHORT ANSWER

50. A newly isolated natural product was shown to be optically active. If a solution of 2.0 g in 10 mL of ethanol in a 50 cm tube gives a rotation of +2.57°, what is the specific rotation of this natural product?

    Answer: +2.57° Section: 6

51. Steroisomers which are not mirror image isomers are _____ .

    Answer: disatereomers   Section: 8

MULTIPLE CHOICE

52. A mixture of equal amounts of two enantiomers _____ .

    a. is called a racemic mixture
    b. is optically inactive
    c. implies that the enantiomers are meso forms
    d. both a and b
    e. none of the above

    Answer: d   Section: 6

SHORT ANSWER

53. Briefly describe how two enantiomers might be separated.

    Answer:
    The two compounds can be converted to diastereomers, separated based on
    different physical properties of these diastereomers, and subsequently
    returned to their original forms or chromatographic separation using a
    chiral stationary phase.
    Section: 13

54. Which of the following terms best describes the pair of compounds shown:
    enantiomers, diastereomers, or the same compound?

    Answer: the same compound   Section: 8

55. Which of the following terms best describes the pair of compounds shown:
    enantiomers, diastereomers, or the same compound?

    Answer: the same compound   Section: 8

56. Which of the following terms best describes the pair of compounds shown: enantiomers, diastereomers, or the same compound?

Answer: enantiomers Section: 8

57. Label each chiral carbon in the compound below as R or S.

```
       CHO
       |
HO─────┼─ H
H──────┼─ OH
H──────┼─ OH
       |
      CH₂OH
```

Answer:

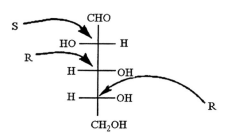

Section: 10

58. Draw the structure of (S)-1-bromo-1-chloropropane. Take particular care to indicate three-dimensional sterochemical detail properly.

Answer:

Section: 5

59. How many chiral carbons are present in the compound below?

Answer: 5 Section: 3

MULTIPLE CHOICE

60. Which of the following statements correctly describes the molecule shown below?

    a. It is achiral.
    b. It is meso.
    c. It's chirality center possesses the R configuration.
    d. The mirror image of this molecule is its enantiomer.
    e. The molecule possesses enantiotopic hydrogens.

Answer: d   Difficulty: 2 Section: 3

SHORT ANSWER

61. The compound below contains a chirality center at nitrogen. Why can't individual stereoisomers of this compound be isolated at room temperature?

Answer:
Nitrogen has a pair of nonbonding electrons that allow it to turn "inside out" at room temperature. Because of rapid inversion, the individual stereoisomers cannot be isolated.

Difficulty: 3 Section: 1

MULTIPLE CHOICE

62. How many chirality centers are present in a molecule of 2,4,6-trimethylheptane?

    a. 0
    b. 1
    c. 2
    d. 3
    e. 4

    Answer: a   Difficulty: 2 Section: 3

63. Which of the following is a true statement?

    a. All chiral molecules possess a plane of symmetry.
    b. All achiral molecules are meso.
    c. All molecules which possess a single chirality center of the S configuration are levorotatory.
    d. A mixture of achiral compounds will be optically inactive.
    e. All molecules which possess 2 or more chirality centers will be chiral.

    Answer: d   Difficulty: 2 Section: 6

64. (-)-Mandelic acid has a specific rotation of -158°. What would be the specific rotation of a solution which contains 40% (-)-mandelic acid and 60% (+)-mandelic acid?
    a. +95°
    b. +63°
    c. +32°
    d. -32°
    e. -63°

    Answer: c   Difficulty: 3 Section: 7

SHORT ANSWER

65. The compound $CH_3CHClCH_2CH=CHCH_2CH_3$ contains only one chirality center. Explain why it has four stereoisomers.

    Answer:
    In addition to the single chirality center which may be either R or S, it also contains a carbon-carbon double bond which mat be either E or Z. Thus, the four stereoisomers can be designated E,R; E,S; Z,R; and Z,S.
    Difficulty: 2 Section: 8

66. Draw the stereoisomers of 1,3-dichlorocyclopentane.

Answer:

Difficulty: 2 Section: 8

67. Draw a perspective formula of (2R,3S)-3-bromo-2-butanol.

Answer:

Difficulty: 2 Section: 10

MULTIPLE CHOICE

68. What term properly describes the stereochemical relationship between hydrogens H_a and H_b in the compound below?

a. homotopic
b. enantiotopic
c. diastereotopic
d. meso
e. syn

Answer: a Difficulty: 3 Section: 14

SHORT ANSWER

69. Provide the structure of the major organic product of each of the following reactions.

Answer:

(racemic)   (meso)

Difficulty: 2 Section: 17

Chapter 5: Reactions of Alkynes, Introduction to Multistep Synthesis

MULTIPLE CHOICE

1. What is the hybridization of the carbon atoms numbered 1 and 2 respectively in the following structure?

   a. sp^3, sp^2
   b. sp^2, sp^2
   c. sp_3, sp
   d. sp^2, sp
   e. sp, sp^2

   $$\langle\!\!\bigcirc\!\!\rangle\!-\!CH\!=\!CH\!-\!\underset{1}{C}\!\equiv\!\underset{2}{CH}$$

   Answer: d   Section: 3

2. Which of the following statements is not true about propyne, $HC\equiv C-CH_3$?

   a. It contains six sigma bonds.
   b. It contains three pi bonds.
   c. The H—C—H bond angle is about 109.5°.
   d. The C—C—C bond angle is 180°.
   e. The pi bond is weaker than the sigma bond.

   Answer: b   Section: 3

3. What is the IUPAC name for the following alkyne?

   $$CH_3-C\equiv C-CH_2-\overset{\overset{\displaystyle Br}{|}}{\underset{\underset{\displaystyle CH_3}{|}}{C}}-CH_3$$

   a. 5-Bromo-2-heptyne
   b. 3-Bromo-5-heptyne
   c. 2-Bromo-2-methyl-4-hexyne
   d. 5-Bromo-5,5-dimethylhexyne
   e. 5-Bromo-5-methyl-2-hexyne

   Answer: e   Section: 1

4. Which is the correct order of decreasing acidity in the following compounds?

H_2O CH_3CH_3 NH_3 $CH_2=CH_2$ $HC\equiv CH$
A   B   C   D   E

a. A > E > C > D > B
b. A > E > D > B > C
c. E > A > C > B > D
d. A > C > E > D > B
e. E > D > B > A > C

Answer: a   Section: 9

5. What is the major product of the following reaction?

$$CH_3C\equiv CH \xrightarrow[\text{excess}]{HCl} \ ?$$

a. CH_3CH_2CH with Cl and Cl

b.
```
        Cl
        |
CH3 — C — CH3
        |
        Cl
```

c. $CH_3CH=CH-Cl$

d. $CH_3C=CH_2$ with Cl

e. $CH_3C=CH$ with Cl Cl

Answer: b   Section: 5

6. Which of the following is the best synthesis of 2,2-dibromopropane

$$CH_3 - \underset{\underset{Br}{|}}{\overset{\overset{Br}{|}}{C}} - CH_3 \quad ?$$

a.
$$CH_3CH= CH_2 \xrightarrow[CCl_4]{Br_2}$$

b.
$$CH_3CH= CH_2 \xrightarrow[light]{Br_2}$$

c.
$$CH_3C\equiv CH \xrightarrow{2HBr}$$

d.
$$CH_3C\equiv CH \xrightarrow[peroxide]{2\ HBr}$$

e.
$$CH_3CH= CH_2 \xrightarrow[H_2O]{Br_2}$$

Answer: c Section: 5

7. Which of the following is the best synthesis of 1,1-dibromopropane,

$$CH_3CH_2\underset{\underset{Br}{|}}{\overset{\overset{Br}{|}}{CH}} \quad ?$$

a.
$$CH_3CH= CH_2 \xrightarrow[CCl_4]{Br_2}$$

b.
$$CH_3CH= CH_2 \xrightarrow[light]{Br_2}$$

c.
$$CH_3C\equiv CH \xrightarrow{2HBr}$$

d.
$$CH_3C\equiv CH \xrightarrow[peroxide]{2\ HBr}$$

e.
$$CH_3CH= CH_2 \xrightarrow[H_2O]{Br_2}$$

Answer: d Section: 5

8. Which of the following are enol forms of 2-butanone,

Answer: e Section: 6

9. Which of the following is the final and major product of this reaction?

   a. I
   b. II
   c. III
   d. IV
   e. V

Answer: a Section: 6

10. Which of the following is the final and major product of this reaction?

a. I
b. II
c. III
d. IV
e. V

I.

II.

III.

IV.

V.

Answer: c Section: 7

11. The reagent needed to convert 2-butyne to cis-2-butene is:

a. H_2/Pt
b. H_2/Lindlar's catalyst
c. Li/NH_3
d. Na/NH_3
e. H^+/Zinc dust

Answer: b Section: 8

12. The reagent needed to convert 2-butyne to trans-2-butene is:

a. H_2/Pt
b. H_2/Lindlar's catalyst
c. Li/NH_3
d. Na/NH_3
e. c or d

Answer: e Section: 8

13. What are the products of the following reaction,

$$CH_3OH + CH_3C \equiv C^- Na^+ \longrightarrow ?$$

a. $CH_3C \equiv CCH_3 + NaOH$
b. $CH_3C \equiv COCH_3 + NaOH$
c. $CH_3C \equiv CH + CH_3O^- Na^+$
d. $CH_3OC \equiv CH + NaCH_3$
e. No reaction

Answer: c Section: 9

14. What is/are the major organic product(s) of the following reaction,

$$HC \equiv C : =CH_3CH_2Br \longrightarrow ?$$

a. $CH_2=CH_2 + HC \equiv CH$
b. $CH_3CH_2C \equiv CH$
c. $HC \equiv CBr$
d. $HC \equiv CCH_2CH_2Br$
e. $CH_3C \equiv CCH_3$

Answer: b Section: 10

SHORT ANSWER

15. Write structures and give IUPAC names for all <u>alkynes</u> with molecular formula C_5H_8.

Answer:

		CH_3
		$\mid$
$CH_3CH_2CH_2C \equiv CH$	$CH_3CH_2C \equiv CCH_3$	$CH_3CHC \equiv CH$
1-pentyne	2-pentyne	3-methyl-butyne

Section: 1

16. Complete the following tree of reactions by giving the major products.

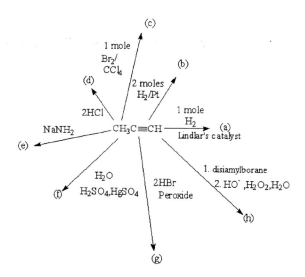

Answer:

a. $CH_3CH{=}CH_2$

b. $CH_3CH_2CH_3$

c.
```
   Br        H
    \       /
     C  =  C
    /       \
  CH3        Br
      trans
```

d.
```
        Cl
        |
CH3 —— C —— CH3
        |
        Cl
```

e. $CH_3C{\equiv}C^-Na^+ + NH_3$

f.
```
CH3CCH3
   ||
   O
```

g.
```
        Br
        |
CH3CH2CH
        |
        Br
```

h.
```
CH3CH2CH
        ||
        O
```

Section: 9

17. Suggest a plausible scheme for the following synthesis

$$CH_3CH_2\underset{\underset{Br}{|}}{C}HCH_3 \quad \text{from} \quad HC\equiv CH$$

Answer:

$$HC\equiv CH + NaNH_2 \longrightarrow HC\equiv C\overset{..}{:} \xrightarrow{CH_3 I} HC\equiv CCH_3 \xrightarrow{NaNH_2} \overset{..}{:}C\equiv CCH_3$$

$$\overset{CH_3 I}{\Big\downarrow}$$

$$1 \text{ mole}$$

$$CH_3CH_2\underset{\underset{\underset{product}{Br}}{|}}{C}HCH_3 \xleftarrow{HBr} CH_3CH\equiv CHCH_3 \xleftarrow[\text{Lindlar's catalyst}]{H_2} CH_2C\equiv CCH_3$$

Section: 11

18. Explain why $HC\equiv N$ is a stronger acid than $HC\equiv CH$.

Answer: Because nitrogen is more electronegative than carbon. Section: 9

19. An optically active compound (A), C_6H_{10}, reacts with H_2/Ni to produce compound (B), C_6H_{14}. (B) is optically inactive. Deduce the structures of (A) and (B).

Answer:
a.
$$HC\equiv C\underset{\underset{CH_3}{|}}{C}HCH_2CH_3$$
optically active

b.
$$CH_3CH_2\underset{\underset{CH_3}{|}}{C}HCH_2CH_3$$
optically inactive

Section: 8

20. Although you might expect 1-propyne to be more reactive than 1-propene in electrophilic addition reactions, the reverse is true. Explain this using your knowledge of the mechanism of electrophilic additions to both alkynes and alkenes.

Answer:
When HCl is added to an alkene, a secondary carbocation intermediate is formed, $CH_3\overset{\oplus}{C}HCH_3$. When HCl is added to an alkyne, a vinylic cation is formed as an intermediate, $CH_3\overset{\oplus}{C}=CH_2$. Alkynes are less stable than alkenes but vinylic carbocations are less stable than secondary carbocations. Thus, alkenes are more reactive than alkynes since the difference in stabilities between the carbocations is greater than the difference in stabilities between the alkene and alkyne.
Section: 4

21. Which hydrogens are more acidic, those of benzene or those of cyclohexane? Why?

benzene   cyclohexane

Answer:
Benzene. The carbon hybridization of benzene is sp^2 while that of cyclohexane is sp^3. The more s character in benzene causes its hydrogens to be more acidic than those of cyclohexane.
Section: 9

22. Give the IUPAC name for $(CH_3)_2C(CH_2CH_3)CCCH(CH_3)_2$.

Answer: 2,5,5-trimethyl-3-heptyne   Section: 1

23. Give the IUPAC name for $HCCCH_2CH_2CH_3$.

Answer: 1-pentyne   Section: 1

24. Give the IUPAC name for $BrCH_2CH_2CCCH_2CH_3$.

Answer: 1-bromo-3-hexyne   Section: 1

25. Draw an acceptable structure for acetylene.

Answer: $H-C\equiv C-H$ Section: 1

26. Draw an acceptable structure for 3-sec-butyl-1-heptyne.

Answer:

Section: 1

27. Draw an acceptable structure for 2-hexyne.

Answer:

Section: 1

MULTIPLE CHOICE

28. How many distinct terminal alkynes exist with a molecular formula of C_5H_8?

   a. 1
   b. 2
   c. 3
   d. 4
   e. 5

   Answer: b Section: 1

29. The carbon-carbon triple bond of an alkyne is composed of _____.

   a. three σ bonds
   b. two σ bonds and one π bond
   c. one σ bond and two π bonds
   d. three π bonds

   Answer: c Section: 3

30. Which of the following <u>improperly</u> describes the physical properties of an alkyne?

   a. relatively nonpolar
   b. nearly insoluble in water
   c. less dense than water
   d. insoluble in most organic solvents

   Answer: d Section: 2

SHORT ANSWER

31. Why are terminal alkynes more acidic than other hydrocarbons?

    Answer:
    The carbanion which results upon deprotonation of a terminal alkyne has the lone pair of electrons in an sp hybrid orbital. The greater % s character of this orbital gives the orbital a significantly lower energy.

    Section: 9

32. Provide the structure of the major organic product(s) in the reaction below.

    $$CH_3CH_2{\equiv}CH \xrightarrow[\text{2. } PhCH_2Br]{\text{1. } NaNH_2}$$

    Answer: $CH_3CH_2{\equiv}CCH_2Ph$ Section: 10

MULTIPLE CHOICE

33. Which of the species below is less basic than acetylide?

    a. CH_3Li
    b. CH_3ONa
    c. $NaOH$
    d. both b and c
    e. all of the above

    Answer: d Section: 9

34. A mixture of 1-heptyne, 2-heptyne, and 3-heptyne was hydrogenated in the presence of a platinum catalyst until hydrogen uptake ceased. If one assumes that the hydrogenation went to completion, how many <u>different</u> seven-carbon hydrocarbons were produced?

    a. 1
    b. 2
    c. 3
    d. 6
    e. 8

    Answer: a Section: 8

SHORT ANSWER

35. Provide the structure of the major organic product(s) in the reaction below.

Answer:

Section: 8

MULTIPLE CHOICE

36. In the reduction of alkynes using sodium in liquid ammonia, which of the species below is <u>not</u> believed to be an intermediate in the commonly accepted mechanism?

    a. vinyl anion
    b. vinyl radical
    c. radical anion
    d. vinyl cation

Answer: d Section: 8

SHORT ANSWER

37. Provide the structure of the major organic product(s) in the reaction below.

$CH_3CH_2-C{\equiv}C-CH_3$ $\xrightarrow{\text{Na, NH}_3}$

Answer:

CH_3CH_2 \ / H
        C=C
H / \ CH_3

Section: 8

38. Provide the structure of the major organic product(s) in the reaction below.

Answer:

Section: 5

MULTIPLE CHOICE

39. In the addition of hydrogen bromide to alkynes in the absence of peroxides, which of the following species is believed to be an intermediate?

  a. vinyl anion
  b. vinyl cation
  c. vinyl radical
  d. carbene
  e. none of the above

Answer: b   Section: 5

SHORT ANSWER

40. Provide the structure of the major organic product(s) in the reaction below.

Answer:

Section: 6

41. Provide the structure of the major organic product(s) in the reaction below.

$$\text{(cyclobutyl)}-C\equiv C-H \xrightarrow[\text{2. } H_2O_2, \,^-OH]{\text{1. disiamylborane}}$$

Answer:

(cyclobutyl)CH$_2$CHO

Section: 7

MULTIPLE CHOICE

42. Among the compounds water, 1-butyne, 2-butyne, and ethane, which are stronger acids than ammonia?

a. 1-butyne and ethane
b. water and 1-butyne
c. water and ethane
d. 1-butyne and 2-butyne

Answer: b Section: 9

43. _____ is produced when 1 equivalent of HBr is added to 1-hexyne in the presence of peroxides.

a. 2-Bromo-1-hexene
b. E-1-Bromo-1-hexene
c. Z-1-Bromo-1-hexene
d. A mixture of E and Z isomers of 1-bromo-1-hexene
e. E-2-Bromo-2-hexene

Answer: d Section: 5

44. Which of the alkyne addition reactions below involves an enol intermediate?

a. hydroboration/oxidation
b. treatment with HgSO$_4$ in dilute H$_2$SO$_4$
c. hydrogenation
d. both a and b
e. none of the above

Answer: d Section: 6

SHORT ANSWER

45. Describe a sequence of reactions by which 3-heptyne can be straightforwardly prepared from acetylene.

    Answer:
    1. $NaNH_2$
    2. CH_3CH_2Br
    3. $NaNH_2$
    4. $CH_3CH_2CH_2Br$
    Section: 11

46. Describe a sequence of reactions by which meso-2,3-dibromobutane can be straightforwardly prepared from propyne.

    Answer:
    1. $NaNH_2$
    2. CH_3Br
    3. Na, NH_3
    4. Br_2, CCl_4
    Section: 11

47. Describe a sequence of reactions by which butylbenzene can be straightforwardly prepared from phenylacetylene.

    Answer:
    1. $NaNH_2$
    2. CH_3CH_2Br
    3. H_2, Pt
    Section: 11

48. Describe a sequence of reactions by which $CH_3CH_2CH_2COCH_2CH_3$ can be straightforwardly prepared from 1-butyne.

    Answer:
    1. $NaNH_2$
    2. CH_3CH_2Br
    3. $HgSO_4, H_2O, H_2SO_4$
    Section: 11

49. Describe a sequence of reactions by which <u>trans</u>-2-pentene can be straightforwardly prepared from propyne.

Answer:
1. $NaNH_2$
2. CH_3CH_2Br
3. Na, NH_3

Section: 11

MULTIPLE CHOICE

50. What are the hybridizations of the carbon atoms numbered 1 and 2 in the structure below?

$$\overset{1}{\underset{}{}}\quad\overset{2}{\underset{}{}}$$
$$H_3C-CH= CH-C \equiv C-H$$

a. sp^3, sp^2
b. sp^2, sp^2
c. sp^2, sp
d. sp, sp

Answer: d Section: 3

SHORT ANSWER

51. For the reaction below indicate whether the equilbrium constant will be greater than 1 or less than 1.

$$CH_3Li + CH_3C\equiv CH \rightleftharpoons CH_4 + CH_3C\equiv CLi$$

Answer: >1 Section: 9

52. For the reaction below indicate whether the equilbrium constant will be greater than 1 or less than 1.

$$CH_3CH_2CHO \rightleftharpoons CH_3CH=CHOH$$

Answer: <1 Section: 6

53. For the reaction below indicate whether the equilbrium constant will be greater than 1 or less than 1.

$$KOH + CH_3C\equiv CH \rightleftharpoons H_2O + CH_3C\equiv CK$$

Answer: <1 Section: 9

54. Complete the following reaction

Answer: ethynylcyclobutane Section: 7

55. Complete the following reaction.

$$\text{3-hexyne} \xrightarrow{\text{Na, NH}_3}$$

Answer: trans-3-heptene Section: 8

MULTIPLE CHOICE

56. How many distinct internal alkynes exist with a molecular formula of C_6H_{10}?

a. 1
b. 2
c. 3
d. 4
e. 5

Answer: c Section: 1

57. What is the common name for the following alkyne?

$$CH_3-\underset{\underset{CH_3}{|}}{\overset{\overset{CH_3}{|}}{C}}-C\equiv CH$$

a. Neohexyne
b. Trimethylpropyne
c. Trimethylacetylene
d. tert-Butylacetylene
e. Isopropylacetylene

Answer: d Section: 1

SHORT ANSWER

58. Give the systematic name for the alkyne $Cl_3CCH_2CH_2CCCH_3$.

Answer: 6,6,6-trichloro-2-hexyne Difficulty: 2 Section: 1

59. Provide the structure of all distinct terminal alkynes with a molecular formula of C_6H_{10}.

Answer: Difficulty: 3 Section: 1

MULTIPLE CHOICE

60. What two atomic orbitals' or hybrid atomic orbitals overlap to form the carbon-carbon σ bond in ethyne?

   a. sp^3-sp^3
   b. p—p
   c. sp^2-sp^2
   d. s—s
   e. sp—sp

Answer: e   Difficulty: 2 Section: 3

61. Which of the following statements correctly describes the general reactivity of alkynes?

   a. An alkyne is an electron-rich molecule and therefore reacts as a nucleophile.
   b. The σ bonds of alkynes are higher in energy than the π bonds and are thus more reactive.
   c. Unlike alkenes, alkynes fail to undergo electrophilic addition reactions.
   d. Alkynes are generally more reactive than alkenes.
   e. None of the above.

Answer: a   Difficulty: 1 Section: 4

62. Which of the following compounds is the major product when 1-hexyne is treated with excess HBr?

   a. 1,1-dibromohexane
   b. 1,1-dibromohexene
   c. 1,2-dibromohexene
   d. 1,2-dibromohexane
   e. 2,2-dibromohexane

Answer: e   Difficulty: 1 Section: 5

SHORT ANSWER

63. Provide the structure of the intermediate which forms in the first step of the reaction of 1-butyne with water in the presence of $H_2SO_4/HgSO_4$.

    Answer:

$$CH_2CH_3C\overset{\overset{\displaystyle Hg^{2+}}{\diagup\diagdown}}{=\!=}CH$$

    Difficulty: 3 Section: 6

64. Describe a sequence of reactions by which the aldehyde $CH_3CH_2CH_2CH_2CHO$ could be prepared from acetylene.

    Answer:
    1. $NaNH_2$
    2. $CH_3CH_2CH_2Br$
    3. disiamylborane
    4. H_2O_2, ^-OH
    Difficulty: 3 Section: 11

65. Describe a sequence of reactions by which <u>cis</u>-2-pentene could be prepared from acetylene.

    Answer:
    1. $NaNH_2$
    2. CH_3CH_2Br
    3. $NaNH_2$
    4. CH_3Br
    5. H_2, Lindlar's catalyst
    Difficulty: 3 Section: 11

MULTIPLE CHOICE

66. Most of the ethyne produced commercially is used as
    _____.

    a. fuel in welding torches
    b. a ripening agent in agricultural warehouses
    c. solvent
    d. a starting material for polymers
    e. refrigerant

    Answer: d   Difficulty: 1 Section: 12

67. What type of organic product results when 1-butyne is treated with 6M aqueous NaOH?

    a. a sodium acetylide
    b. an enol
    c. an alcohol
    d. an aldehyde
    e. not much reaction; most of the alkyne remains unchanged

    Answer: e   Difficulty: 3 Section: 9

MULTIPLE CHOICE

1. Which of the following pairs are resonance structures?

    a. I
    b. II
    c. III
    d. IV
    e. V

I.

and

II.

and

III.

and

IV.

and

V.

and

Answer: c Section: 3

2. Which of the following pairs are resonance structures?

    a. I
    b. II
    c. III
    d. IV
    e. V

    I. $CH_2=CHCH_3$ and   △

    II. $CH_2{=}C{-}H$ and $CH_3{-}C{-}H$
             :ÖH   :Ö:

    III. $CH_3{-}C{-}H$ and $CH_3{-}\overset{\oplus}{C}{-}H$
              :Ö: :Ö:$_\ominus$

    IV. $CH_3{-}C{-}H$ and $CH_3{-}\overset{\ominus}{C}{-}H$
               Ö   :Ö:$_\oplus$

    V. $CH_3{-}\overset{..}{\underset{..}{O}}{-}CH_3$ and $CH_3CH_2\overset{..}{\underset{..}{O}}H$

    Answer: c Section: 3

3. Which of the following statements is incorrect about benzene?

    a. All of the carbon atoms are sp hybridized.
    b. It has delocalized electrons.
    c. The carbon-carbon bond lengths are all the same.
    d. The carbon-hydrogen bond lengths are all the same.
    e. All twelve atoms lie in the same plane.

    Answer: a Section: 2

4. Which of the following is a resonance structure of the species shown?

    a. I
    b. II
    c. III
    d. IV
    e. V

Answer: e Section: 3

5. Which of the following statements about benzene is/are correct?

    a. All of the carbon atoms are sp^3 hybridized.
    b. It has no delocalized electrons.
    c. The carbon-carbon bond length is longer than that of ethane.
    d. It is a planar molecule.
    e. The carbon-hydrogen bonds are not the same length.

Answer: d Section: 2

6. Which of the following is the most stable resonance contributor to acetic acid?

$$CH_3COH$$

a.

CH$_3$— C $=$ O— H
        $\ominus$ $\oplus$
        ‖
       :O:

b.

CH$_3$— C $—$ O$=$H
        ‖ $\oplus$ $\ominus$
       :O:

c. $\ominus$

CH$_3$$=$ C— O— H
        |
       :O:

d.

CH$_2$$=$ C— O— H
        $\oplus$
        |
       :OH

e.

CH$_3$— C— O— H
        $\oplus$
        |
       :O:
        $\ominus$

Answer: e Section: 5

117

7. Which of the following is an allylic cation?

    a. I
    b. II
    c. III
    d. IV
    e. V

    I. $CH_3\overset{\oplus}{C}H=CH_2$

    II. $CH_3CH=CH-\overset{\oplus}{C}H_2$

    III. $CH_3\overset{\oplus}{C}HCH_3$

    IV. $CH_3\overset{\oplus}{C}HCH_2CH=CH_2$

    V.

Answer: b Section: 7

8. Which of the following is a benzylic cation?

    a. I
    b. II
    c. III
    d. IV
    e. V

    I.

    II.

    III.

    IV.

    V.

Answer: d Section: 7

118

9. Which of the following is the most stable cation?

    a. I
    b. II
    c. III
    d. IV
    e. V

I. $CH_3\overset{\oplus}{C}=CH_2$

II. $CH_3CH=\overset{\oplus}{C}-$⬡

III. $CH_3-\underset{\underset{\oplus}{}}{\overset{\overset{CH_3}{|}}{C}}-$⬡

IV. $CH_3-\underset{\underset{CH_3}{|}}{\overset{\oplus}{C}}-CH_3$

V.   ⬡$-\overset{\oplus}{C}H-CH_3$

Answer: c Section: 7

119

10. Which of the following is the <u>most</u> stable radical?

   a. I
   b. II
   c. III
   d. IV
   e. V

   I.   ĊHCH₃

   II.   CH₂–ĊH₂

   III. CH₂CH₃

   IV.   CH₂CH₃

   V.   CH₂CH₃

   Answer: a Section: 8

11. How many allylic hydrogen atoms are present in the following molecule?

   a. 2
   b. 3
   c. 4
   d. 5
   e. zero

   $CH_2=CH—CH=CH$ —

   Answer: e Section: 7

12. How many benzylic hydrogens are present in the following molecule?

    a. 2
    b. 3
    c. 4
    d. 5
    e. zero

CH_2=CH—CH=CH—

Answer: e Section: 7

13. Which of the following is/are the main product(s) of the following reaction?

    $$CH_2=CH-CH=CH_2 \xrightarrow{HBr} \quad ?$$

    a. $CH_2CH=CHCH_3$
       |
       Br
    b. $CH_2CH_2CH=CH_2$
       |
       Br
    c. $CH_2=CHCHCH_3$
            |
            Br
    d. a and b
    e. a and c

Answer: e Section: 9

14. Which of the following is/are the major product(s) of the following reaction?

    a. I.
    b. I. and II.
    c. I. and III.
    d. II.
    e. III.

Answer: a Section: 9

15. Which of the following is/are the <u>major</u> product(s) of the following reaction?

    a. I
    b. I and II
    c. II
    d. II and III
    e. III

Answer: c Section: 9

16. Which of the following is the <u>strongest</u> acid?

    a. CH_3CH_2OH
    b. CH_3OCH_3
    c. CH_3CH
           $\|$
          O

    d. CH_3COCH_3
            $\|$
          O

    e. CH_3COH
           $\|$
         O

Answer: e Section: 10

17. Which of the following is the <u>strongest</u> acid?

    a. I
    b. II
    c. III
    d. IV
    e. V

I.

II.

III.

IV.

V.

Answer: d Section: 10

18. In the molecular orbital representation of benzene, how many π molecular orbitals are present?

    a. 1
    b. 2
    c. 4
    d. 6
    e. 8

    Answer: d Section: 11

19. Which species is represented by the following distribution of π electrons in the molecular orbitals and energy levels diagram?

    a. I
    b. II
    c. III
    d. IV
    e. V

I.

II. $CH_2=CH-\overset{\oplus}{C}H_2$

III. $CH_2=CH-\overset{\bullet}{C}H_2$

IV. $CH_2=CH-\overset{\ominus}{C}H_2$

V.

Answer: c Section: 11

SHORT ANSWER

20. Draw two other resonance contributors for the following compound:

    Answer:

      and

    Section: 4

21. Are the following pairs of structures resonance contributors or different compounds?

 and

Answer: They are resonance contributors. Section: 4

22. Draw four resonance contributors for the following structure and indicate which is the <u>most</u> important contributor. Explain.

$$\left[CH_2\!=\!\!N\!\!=\!\!O \atop \quad\;\;\overset{O}{\|} \right]^{\ominus}$$

a. $\overset{\oplus}{CH_2}\!-\!\overset{\overset{:\ddot{O}:^\ominus}{|}}{\underset{}{N}}\!-\!\ddot{\ddot{O}}:^\ominus$

b. $\overset{\ominus}{CH_2}\!-\!\overset{\overset{\ddot{O}:}{\|}}{\underset{\oplus}{N}}\!-\!\ddot{\ddot{O}}:^\ominus$

c. $CH_2\!=\!\overset{\overset{:\ddot{O}:^\ominus}{|}}{\underset{\oplus}{N}}\!-\!\ddot{\ddot{O}}:^\ominus$

d. $\overset{\ominus}{CH_2}\!-\!\overset{\overset{:\ddot{O}:^\ominus}{|}}{\underset{\oplus}{N}}\!=\!\ddot{O}:$

Answer:
C is the most important contributor because all the atoms have complete octets and the negative charges are on the most electronegative atoms.
Section: 5

23. Which of the following numbered hydrogens is most easily abstracted by heterolytic cleavage? Explain.

$$\underset{\underset{1}{H}}{\overset{H}{\diagdown}} C\!=\!\underset{\overset{2}{H}}{\overset{}{C}}\!-\!\underset{H}{\overset{H}{\underset{|}{\overset{|}{C}}}}\!-\!H_3$$

Answer:
Hydrogen #3 is most easily abstracted by heterolytic cleavage because the carbocation produced from the cleavage is allylic.
Section: 7

24. Rate the following molecules in decreasing order of stability:

a   b   c

Answer:
a > c > b
aromatic > linear > antiaromatic.
Section: 5

25. Which of the following compounds is more stable? Explain.

a. $\overset{\oplus}{C}H_2$ or   b. $\overset{\oplus}{C}HCH_3$

Answer:
(a) and (b) are both benzylic cations. (b) is more stable because it is more alkyl substituted.
Section: 7

26. Which of the following compounds is more stable? Explain.

a. $\overset{CH_3}{\underset{\cdot}{C}}-CH_3$ or   b. $\overset{CH_3}{\underset{\cdot}{C}}-H$

Answer:
(a) is more stable than (b). Although both species are benzylic, (a) is more alkyl substituted than (b), making it more stable.
Section: 8

27. Why is phenol a stronger acid than cyclohexanol?

OH   OH

phenol   cyclohexanol

Answer:
The anion of phenol is stabilized by resonance, causing the proton to be more easily abstracted than that of cyclohexanol.
Section: 10

28. Draw the important resonance structures of:

$H_2\bar{C}-\overset{+}{N}\equiv N\!:$

Answer:

$H_2\bar{C}-\overset{+}{N}\equiv N$ ⟷ $H_2C=\overset{+}{N}=\ddot{N}\!:^-$

Section: 4

29. Draw the important resonance structures of:

$CH_3CH=\overset{+}{C}HCHCH_2CH_3$

Answer:

$CH_3CH=\overset{+}{C}HCHCH_2CH_3$ and $CH_3\overset{+}{C}HCH=CHCH_2CH_3$

Section: 4

30. Draw the important resonance structures of:

Answer:

Section: 4

31. Draw the important resonance contributing forms for the structure hown below.

Answer:

Section: 4

32. Draw the important resonance contributing forms for the structure shown below.

Answer:

Section: 4

33. Draw the important resonance contributing forms for the structure shown below.

Answer:

Section: 4

34. Stabilization of a charged species usually results when this species can be more accurately depicted as a hybrid of several resonance contributing forms. Why is this the case?

Answer: Stabilization results from delocalization of charge.

Section: 5

35. What is suggested by the fact that benzene's molar heat of hydrogenation is 36 kcal less than three times the molar heat of hydrogenation of cyclohexene?

Answer:
This suggests that the type of π bonding in benzene lends special stability to the molecule.

Section: 6

36. Provide a diagram which depicts the relative energies of the π molecular orbitals of benzene. Show which molecular orbitals are filled in benezene's ground state.

Answer:

Section: 11

MULTIPLE CHOICE

37. In which of the following does resonance delocalization of electron density <u>not</u> play a role?

   a. allyl cation
   b. benzyl anion
   c. CO_3^{2-}
   d. O_3
   e. cyclohexyl radical

Answer: e   Difficulty: 3 Section: 5

SHORT ANSWER

38. In the acetate ion ($CH_3CO_2^-$), which of the oxygen atoms bears a greater negative charge? Explain your answer.

Answer:
The acetate ion is a hybrid of the two resonance contributors below. Each oxygen in this ion has the same charge (-1/2); the oxygen atoms are indistinguishable and the carbon-oxygen bonds are identical.

Difficulty: 2 Section: 5

MULTIPLE CHOICE

39. Which of the following statements concerning resonance contributors and resonance hybrids is <u>not</u> correct?

     a. The fewer nearly equivalent resonance contributors are in the structure, the greater the resonance energy.
     b. A resonance hybrid is more stable than the predicted stability of any of its resonance contributors.
     c. The greater the number of relatively stable resonance contributors, the greater the resonance energy.
     d. The greater the predicted stability of a resonance contributor, the more it contributes to the resonance hybrid.
     e. None of the above.

Answer: a   Difficulty: 2   Section: 6

SHORT ANSWER

40. Which is more stable, a tertiary alkyl radical or a benzylic radical? Explain.

Answer:
A benzylic radical is more stable than an alkyl radical since the unpaired electron can be delocalized by resonance.
Difficulty: 2   Section: 8

41. Provide the structure of the major organic product(s) in the reaction below.

Answer:

Difficulty: 2   Section: 9

42. Provide the structure of the major organic product(s) in the reaction below.

Answer:

+ enantiomer

Difficulty: 2 Section: 9

MULTIPLE CHOICE

43. Which of the following ions is most acidic?

    a. $C_6H_5NH_3^+$
    b. $C_6H_5CH_2NH_3^+$
    c. $(CH_3)_2CHNH_3^+$
    d. $CH_3CH_2NH_3^+$
    e. $CH_3NH_3^+$

    Answer: a   Difficulty: 2 Section: 10

44. Which of the following is the weakest base?

    a. phenolate ion $(C_6H_5O^-)$
    b. ethoxide ion $(CH_3CH_2O^-)$
    c. hydroxide ion (HO^-)
    d. acetate ion $(CH_3CO_2^-)$
    e. methoxide ion (CH_3O^-)

    Answer: d   Difficulty: 2 Section: 10

SHORT ANSWER

45. Show how the p orbitals overlap to form the nonbonding π molecular orbital of the allyl system.

    Answer:

Difficulty: 3 Section: 11

MULTIPLE CHOICE

46. Excluding the horizontal node which is coincident with the molecular plane, how many nodes are present in Ψ_4 of the 1,3,5-hexatriene π system?

   a. 6
   b. 5
   c. 4
   d. 3
   e. 0

   Answer: d   Difficulty: 3 Section: 11

Chapter 7:   Reactions of Dienes

MULTIPLE CHOICE

1. Which of the following is a cumulated diene?

    a. I
    b. II
    c. III
    d. IV
    e. V

    I.
    II.
    III.
    IV.
    V.

    Answer: c Section: 98

2. Which of the following is an isolated diene?

    a. I
    b. II
    c. III
    d. IV
    e. V

    I.
    II.
    III.
    IV.
    V.

    Answer: d Section: 98

3. Which of the following is a conjugated diene?

   a. I
   b. II
   c. III
   d. IV
   e. V

   I.

   II.

   III.

   IV.

   V.

Answer: a Section: 98

4. Give the IUPAC name for the following compound.

   a. Isoprene
   b. 3-Methyl-1,3-butadiene
   c. 2-Methyl-1,3-butadiene
   d. 2-Methyl-2,3-butadiene
   e. Isobutadiene

Answer: c Section: 1

5. What is the IUPAC name for the following compound?

   a. cis,trans-2,4-heptadiene
   b. 2Z, 4Z-2,4-heptadiene
   c. cis, cis-2,4-heptadiene
   d. trans,trans-2,4-heptadiene
   e. 2E, 4E-2,4-heptadiene

Answer: a Section: 2

6. Which of the following is the most stable diene?

    a. I
    b. II
    c. III
    d. IV
    e. V

    I.

    II.

    III.

    IV.

    V.

Answer: c Section: 3

7. How many nodes would you expect in the molecular orbital created by the following alignment of overlapping p orbitals in 1,3-butadiene?

    a. zero
    b. one
    c. two
    d. three
    e. four

Answer: c Section: 4

8. What is/are the product(s) from the following reaction?

   a. I. only
   b. II. only
   c. I. is minor, II. is major
   d. I. is major, II. is minor
   e. Equal amounts of I. and II.

Answer: d Section: 6

9. What is/are the product(s) of the following reaction?

$CH_2=CHCH=CH_2$ + HBr $\longrightarrow$?

   a. I. only
   b. II. only
   c. III. only
   d. I. and II.
   e. I. and III.

Answer: e Section: 7

10. What is/are the product(s) from the following reaction?

    a. I. and II.
    b. I. and III.
    c. I. and IV.
    d. II. and III.
    e. II. and IV.

Answer: a Section: 7

11. What is/are the product(s) from the following reaction?

$$CH_2=CHCH=CH_2 + HBr \xrightarrow{-80° C} ?$$

I. $CH_2CHCH=CH_2$
       |
      Br

II. $CH_3CHCH=CH_2$
        |
       Br

III. $CH_3CH=CHCH_2$
          |
         Br

    a. II. Minor, III. Major
    b. II. Major, III. Minor
    c. III. only
    d. I. and II.
    e. II. and III.

Answer: b Section: 7

12. What is/are the product(s) from the following reaction?

$$CH_2=CHCH=CH_2 + HBr \xrightarrow{45°} \ ?$$

I. $CH_2CH_2CH=CH_2$
$\ \ \ \ |$
$\ \ \ \ Br$

II. $CH_3CHCH=CH_2$ III. $CH_3CH=CHCH_2$
$\ \ \ \ \ \ \ \ |$ $|$
$\ \ \ \ \ \ \ \ Br$ Br

a. I. only
b. II. only
c. III. only
d. I. and II.
e. II. and III.

Answer: c Section: 7

13. What is the thermodynamic product for the following reaction?

a. I
b. II
c. III
d. IV
e. V

Answer: d Section: 8

14. What is the kinetic product for the following reaction?

a. I
b. II
c. III
d. IV
e. V

I.

II.

III.

IV.

V.

Answer: e Section: 8

15. What is the product of the following reaction?

  a. I
  b. II
  c. III
  d. IV
  e. V

 + ‖ $\xrightarrow{\text{heat}}$?

I.

II.

III.

IV.

V.

Answer: e Section: 9

16. What is the product from the following reaction?

a. I
b. II
c. III
d. IV
e. V

I.

COOH
COOH

II.

COOH
COOH

III.

COOH
""""COOH

IV.

COOH
""""COOH

V.

COOH
COOH

H   COOH

+

H   COOH

heat
⟶ ?

Answer: b Section: 9

17. Which of the following conjugated dienes would <u>not</u> react with a dieneophile in a Diels-Alder reaction?

    a. I
    b. II
    c. III
    d. IV
    e. V

    I.

    II.

    III.

    IV.

    Answer: e Section: 9

18. Which of the following dienes is the most reactive in a Diels-Alder reaction?

    a. I
    b. II
    c. III
    d. IV
    e. V

    I.

    II.

    III.

    IV.

    V.

    Answer: c Section: 9

19. What diene and what dieneophile would be used to synthesize the following compound?

    a. I
    b. II
    c. III
    d. IV
    e. V

I.

II.

III.

IV.

V.

Answer: b Section: 9

20. What is the name of the following compound?

    a. Spiro [2.2.2] octane
    b. Bicyclo [2.2.0] octane
    c. Bicyclo [2.2] octane
    d. Bicyclo [2.2.2] octane
    e. Tricyclo [2.2.2] octane

Answer: d Section: 10

21. What is the name of the following compound?

    a. Spiro [5.2] octane
    b. Spiro [5.2.0] octane
    c. Bicyclo [5.2.0] octane
    d. Bicyclo [5.2] octane
    e. Spiro [6.4] octane

Answer: a Section: 10

SHORT ANSWER

22. Arrange the following unsaturated hydrocarbons in order of decreasing reactivity toward HBr addition:

    I.   II.   III.   IV.

    Answer: III > II > IV > I Section: 7

23. Name the following compound:

$$
\begin{array}{ccc}
CH_3 & & H \\
\diagdown & / & \\
C&=&C \quad CH_3 \\
/ & \diagdown & / \\
H & & C=C \\
& / & \diagdown \\
& H & H
\end{array}
$$

Answer:
(2E, 4Z)-2,4-hexadiene or trans,cis-2,4-hexadiene or (2Z, 2E)-2, 4-hexadiene or cis, trans-2,4-hexadiene
Section: 2

24. What is the hybridization of the carbons labeled A, B, and C?

$$
\begin{array}{c}
CH_3 \\
|\\
CH_3 - C = CH - CH_2 - C \equiv CH \\
\quad\quad A \quad\quad\quad B \quad\quad\quad C
\end{array}
$$

Answer:
sp^2, sp^3, sp
Section: 98

25. What is the major product obtained from the addition of one mole of HCl at high temperature to 1,3-butadiene?

    Answer: 1-chloro-2-butene (thermodynamic product)   Section: 8

26. What is the product of the following Diels-Alder reaction?

    Answer:

    Section: 9

27. What diene and what dieneophile could be used to prepare the following?

    Answer:

    Section: 9

28. How could the following compound be synthesized using a Diels-Alder reaction?

    Answer:

    Section: 9

29. Cyclopentadiene can react with itself in a Diels-Alder reaction. Draw the endo and exo products.

    Answer:

    endo   exo

    Section: 9

30. Rank the following dienes in order of increasing stability: trans-1,3-pentadiene, cis-1,3-pentadiene, 1,4-pentadiene, and 1,2-pentadiene.

    Answer:
    1,2-pentadiene < 1,4-pentadiene < cis-1,3-pentadiene < trans-1,3-pentadiene
    Section: 3

31. Among a series of isomeric trienes, the more negative the $\Delta H°$ of the hydrogenation reaction of a given triene, the _____ stable it is relative to the others in the isomeric series.

    Answer: less Section: 3

32. Consider the hydrogenation reaction of each compound listed and rank the compounds in order of increasing $\Delta H°$ of this reaction. The most negative H° should be listed first.

    cis-2-pentene, 2,3-pentadiene, and trans-1,3-pentadiene

    Answer: 2,3-pentadiene < trans-1,3-pentadiene < cis-2-pentene Section: 3

33. Consider the hydrogenation reaction of each compound listed and rank the compounds in order of increasing $\Delta H°$ of this reaction. The most negative H° should be listed first.

    2,5-dimethyl-1,3-cycloheptadiene,
    1,4-dimethyl-1,3-cycloheptadiene, and
    3,6-dimethyl-1,4-cycloheptadiene

    Answer:
    3,6-dimethyl-1,4-cycloheptadiene <
    2,5-dimethyl-1,3-cycloheptadiene <
    1,4-dimethyl-1,3-cycloheptadiene
    Section: 3

MULTIPLE CHOICE

34. What is the hybridization of the central carbon of allene
    (1,2-propadiene)?

    a. sp
    b. sp^2
    c. sp^3
    d. p
    e. none of the above

    Answer: a Section: 98

35. What descriptive term is applied to the type of diene represented by
    2,4-hexadiene?

    a. conjugated diene
    b. cumulated diene
    c. isolated diene
    d. alkynyl diene
    e. none of the above

    Answer: a Section: 98

36. What descriptive term is applied to the type of diene represented by
    1,5-octadiene?

    a. conjugated diene
    b. cumulated diene
    c. isolated diene
    d. alkynyl diene
    e. none of the above

    Answer: c Section: 98

SHORT ANSWER

37. Give a representation of the highest occupied π MO of 1,3-butadiene in
    its ground state.

    Answer:

    Section: 4

38. Give a representation of the lowest unoccupied π MO of 1,3-butadiene in its ground state.

    Answer:

    Section: 4

MULTIPLE CHOICE

39. How many nodes, other than the node coincident with the molecular plane, are found in the highest energy π MO of 1,3-butadiene?

    a. 0
    b. 1
    c. 2
    d. 3
    e. none of the above

    Answer: d Section: 4

SHORT ANSWER

40. Draw (Z)-1,3-hexadiene in its s-trans conformation.

    Answer:

    Section: 2

41. Provide the structure of the major product which results from 1,2-addition of HBr to the diene shown below.

Answer:

Section: 7

42. Provide the structure of the major product which results from 1,4-addition of Br_2 to the diene shown below.

Answer:

Section: 7

43. Draw the structure of the major product which results when the diene shown is treated with HBr at 40° C.

CH$_3$

CH$_3$

Answer:

CH$_3$

CH$_3$

CH$_2$Br

CH$_3$

Section: 8

44. Draw the structure of the major product which results when the diene shown is treated with HBr at -80° C.

CH$_3$

CH$_3$

Answer:

CH$_3$

CH$_3$

Br

CH$_3$

Section: 8

45. In the addition of HBr to conjugated dienes, is the product which results from 1,2-addition or that which results from 1,4-addition typically the product of kinetic control?

Answer: The product of kinetic control typically results from 1,2-addition.

Section: 8

MULTIPLE CHOICE

46. Which of the following terms best describes a Diels-Alder reaction?

   a. a [4+2] cycloaddition
   b. a [2+2] cycloaddition
   c. a sigmatropic rearrangement
   d. a 1,3-dipolar cycloaddition
   e. a substitution reaction

   Answer: a   Section: 9

47. When 1,3-butadiene reacts with $CH_2=CHCN$, which of the terms below best describes the product mixture?

   a. a mixture of two diastereomers
   b. a single compound
   c. a racemic mixture
   d. optically active
   e. a mixture of bicyclic compounds

   Answer: c   Section: 9

48. The Diels-Alder reaction is a concerted reaction; this means:

   a. a mixture of endo and exo products is formed.
   b. all bond making and bond breaking occurs simultaneously.
   c. the products contain rings.
   d. the reaction follows Markovnikov's rule.
   e. the reaction is highly endothermic.

   Answer: b   Section: 9

SHORT ANSWER

49. Provide the structure of the major organic product in the following reaction.

Answer:

Section: 9

50. Provide the structure of the major organic product in the following reaction.

Answer:

Section: 9

51. Provide the structure of the major organic product in the following reaction.

Answer:

Section: 9

52. Provide the structure of the major organic product in the following reaction.

Answer:

Section: 9

53. Provide the structure of the major organic product in the following reaction.

Answer:

Section: 9

153

54. Provide the structure of the major organic product in the following reaction.

Answer:

Section: 9

55. What diene and dienophile would react to give the product below?

Answer:

Section: 9

56. What diene and dienophile would react to give the product below?

Answer:

Section: 9

57. What diene and dienophile would react to give the product below?

Answer:

○ + ⌐CN
         ⌐CN

Section: 9

58. What diene and dienophile would react to give the product below?

Answer:

⌐ + ⌐O
         ⌐O
         ⌐O

Section: 9

MULTIPLE CHOICE

59. Which of the following compounds is the most reactive dienophile in a Diels-Alder reaction with 1,3-butadiene?

   a. $CH_2=CHOCH_3$
   b. $CH_2=CHCHO$
   c. $CH_3CH=CHCH_3$
   d. $(CH_3)_2C=CH_2$
   e. $CH_2=CH_2$

Answer: b Section: 9

SHORT ANSWER

60. Why does the (E,E)-isomer of 2,4-hexadiene react much more rapidly with dienophiles than does the (Z,Z)-isomer?

    Answer:
    The s-cis conformation, which is required for participation in a Diels-Alder reaction, is hindered by a severe CH_3/CH_3 steric interaction in the (Z,Z)-isomer.
    Section: 9

61. Give the systematic name of the compound shown below.

    Answer: 1,7-dimethyl-1,3-cycloheptadiene   Difficulty: 2 Section: 1

62. Give the systematic name of the compound shown below.

    Answer: (S)-5-methyl-5-hexen-2-ol   Difficulty: 2 Section: 1

63. Provide the structure of (2E,4E)-2-chloro-2,4-octadiene.

    Answer: Difficulty: 2 Section: 2

MULTIPLE CHOICE

64. Which of the following compounds contains the longest carbon-carbon single bond?

    a. allene
    b. 1,3-butadiyne
    c. 1,3-butadiene-
    d. propyne
    e. ethane

    Answer: e   Difficulty: 3   Section: 3

SHORT ANSWER

65. Show how the p orbitals overlap to form the highest energy π molecular orbital of 1,3-butadiene.

    Answer:

    Difficulty: 2   Section: 4

MULTIPLE CHOICE

66. What compound results from the 1,4-addition of one equivalent of HBr to 1,3-butadiene?

    a. 1-bromo-1-butene
    b. 2-bromo-2-butene
    c. 4-bromo-1-butene
    d. 3-bromo-1-butene
    e. 1-bromo-2-butene

    Answer: e   Difficulty: 3   Section: 7

67. Which of the following is a correct statement considering thermodynamic versus kinetic control of organic reactions?

    a. When a reaction is under thermodynamic control, the relative amounts of products depend on the activation energies of the steps leading to their formation.
    b. The kinetic product always predominates when the reaction is reversible.
    c. Higher temperatures and longer reaction times typically favor the kinetic product over the thermodynamic product.
    d. When the products are in equilibrium under the reaction conditions, the relative amounts of products depend on their stabilities and the reaction is under thermodynamic control.
    e. None of the above.

    Answer: d   Difficulty: 2 Section: 8

SHORT ANSWER

68. Provide the structure of <u>cis</u>-bicyclo[5.3.0]decane.

    Answer:

    Difficulty: 2 Section: 10

MULTIPLE CHOICE

69. Which of the following statements is <u>not</u> true of the Diels-Alder reaction?

    a. When the diene is electron-rich, the reaction works best when the dienophile contains one or more electron-withdrawing groups conjugated to its C=C.
    b. Substituents which are <u>cis</u> in the dienophile remain <u>cis</u> in the product.
    c. Dienes which cannot achieve an s-<u>cis</u> conformation do not react in Diels-Alder reactions.
    d. Secondary orbital interactions typically cause the endo product to be favored kinetically over the exo.
    e. None of the above.

    Answer: e   Difficulty: 3 Section: 9

70. When 1,3-cyclopentadiene reacts with the <u>cis</u>-isomer of NCCH=CHCN, the major product is _____.

    a. optically active
    b. a meso compound
    c. a racemic mixture
    d. a spirocyclic compound
    e. a fused bicyclic compound

    Answer: b   Difficulty: 3   Section: 9

Chapter 8:   Reactions of Alkanes, Radicals

MULTIPLE CHOICE

1. The major type of reactions that alkanes undergo is:

    a. electrophilic substitution reactions.
    b. electrophilic addition reactions.
    c. free radical substitution reactions.
    d. free radical addition reactions.
    e. nucleophilic substitution reactions.

    Answer: c Section: 2

2. Which of the following is the best description of propane, $CH_3CH_2CH_3$, at room temperature?

    a. Liquid, soluble in H_2O
    b. Gas, soluble in gasoline
    c. Liquid, soluble in gasoline
    d. Gas, soluble in water
    e. Solid, soluble in gasoline

    Answer: b Section: 98

3. Which of the following is the <u>most</u> stable radical?

    a. I
    b. II
    c. III
    d. IV
    e. V

    I. $CH_2=\overset{\bullet}{C}H$

    II. $CH_2=CHCH_2\overset{\bullet}{C}H_2$

    III. $CH_3\overset{\bullet}{C}HCH_3$

    IV. $CH_3\overset{\bullet}{C}H_2$

    V. $CH_3\overset{\bullet}{C}H$—⬡

    Answer: e Section: 5

4. Which of the following is the initiation step for the monobromination of cyclohexane?

   a. I
   b. II
   c. III
   d. IV
   e. V

I.

II.

III.

IV.

V.

Answer: d Section: 2

5. Which of the following is the rate-determining step for the monobromination of cyclohexane?

   a. I
   b. II
   c. III
   d. IV
   e. V

I.

II.

III.

IV.

V.

Answer: b Section: 2

6. Calculate the overall $\Delta H°$ for the reaction given below:

$$(CH_3)_3-C-H + Cl-Cl \rightarrow (CH_3)_3-Cl + H-Cl$$
$$(\Delta H° =91) \qquad (\Delta H° =58) \quad (\Delta H° =78.5) \quad (\Delta H° =103)$$

a. +181.5 kcal/mole
b. +58.0 kcal/mole
c. +32.5 kcal/mole
d. -32.5 kcal/mole
e. -57.5 kcal/mole

Answer: d Section: 2

7. How many products are formed from the monochlorination of ethylcyclohexane? Ignore stereoisomers.

a. 6
b. 8
c. 5
d. 9
e. 11

Answer: a Section: 2

8. Which of the following are the products obtained from the disproportionation of the propyl radical?

$$CH_3CH_2\overset{\bullet}{C}H_2$$

a. $CH_3\overset{\bullet}{C}HCH_3 + CH_2{=}CHCH_3$
b. $CH_3CH_2CH_3 + CH_3\overset{\bullet}{C}HCH_3$
c. $CH_3CH_2CH_2CH_2CH_2CH_3$
d. $CH_3CH_2CH_3 + CH_2{=}CHCH_3$
e. $CH_2{=}C{=}CH_2 + CH_3CH_2CH_3$

Answer: d Section: 2

9. Which of the following is true about the halogenation of methane, CH_4?

   a. The activation energy for $F \cdot + CH_4 \rightarrow \cdot CH_3 + HF$ is larger than any other halogen.
   b. O_2 slows these reactions by inhibiting the termination step.
   c. The activation energy for $I \cdot + CH_4 \rightarrow \cdot CH_3 + HI$ is relatively small.
   d. The reaction must take place in the liquid state.
   e. I_2 is unreactive because the activation energy for $I \cdot + CH_4 \rightarrow \cdot CH_3 + HI$ is relatively large.

   Answer: e   Section: 4

10. When (R)-2-bromobutane reacts with Cl_2/hv, which of the following is true?

    $$
    \begin{array}{cc}
    CH_3 & CH_3 \\
    Br-\!\!\!|\!\!-H & Br-\!\!\!|\!\!-H \\
    Cl-\!\!\!|\!\!-H & H-\!\!\!|\!\!-Cl \\
    CH_3 & CH_3 \\
    I & II
    \end{array}
    $$

    a. Only I is formed.
    b. Only II is formed.
    c. I and II are formed in equal amounts.
    d. I and II are formed in unequal amounts.
    e. I and II could never form under these conditions.

    Answer: d   Section: 6

11. When (R)-2-bromobutane reacts with Cl_2/hv, which of the following is true?

    $$
    \begin{array}{cc}
    CH_2Cl & CH_2Cl \\
    Br-\!\!\!|\!\!-H & H-\!\!\!|\!\!-Br \\
    CH_2CH_3 & CH_2CH_3 \\
    I & II
    \end{array}
    $$

    a. I is formed.
    b. II is formed.
    c. I and II are formed in equal amounts.
    d. I and II are formed in unequal amounts.
    e. I and II could never form under these conditions.

    Answer: a   Section: 6

12. When (R)-2-chlorobutane reacts with Br_2/hv, which of the following is true?

$$
\begin{array}{cc}
CH_3 & CH_3 \\
Br-\!\!\!\mid\!\!\!- Cl & Cl-\!\!\!\mid\!\!\!- Br \\
CH_2CH_3 & CH_2CH_3 \\
I & II
\end{array}
$$

a. Only I is formed.
b. Only II is formed.
c. Both I and II are formed in equal amounts.
d. Both I and II are formed in unequal amounts.
e. Both I and II could never form under these conditions.

Answer: c Section: 6

13. How many dichlorinated products can be isolated when (S)-2-chlorobutane reacts with Cl_2/hv?

$$
\begin{array}{c}
CH_3CH_2 \\
Cl-\!\!\!\mid\!\!\!- H \\
CH_3
\end{array}
$$
(S)-2-chlorobutane

a. 1
b. 2
c. 3
d. 4
e. 5

Answer: e Section: 2

14. What is the <u>major</u> product of the following reaction?

    a. I
    b. II
    c. III
    d. IV
    e. V

Answer: e Section: 5

15. What is the <u>major</u> product of the following reaction?

   a. I
   b. II
   c. III
   d. IV
   e. V

Answer: d Section: 5

16. What is the <u>major</u> product of the following reaction?

    a. I
    b. II
    c. III
    d. IV
    e. V

$$\text{cyclopentene-CH}_3 \xrightarrow[\text{Peroxide}]{\text{HBr}} \ ?$$

I.

II.

III.

IV.

V.

Answer: c Section: 5

17. What is the <u>major</u> product of the following reaction?

    a. I
    b. II
    c. III
    d. IV
    e. V

$$\triangle \xrightarrow[\text{FeCl}_3]{\text{Cl}_2} \ ?$$

I.

II.

III. $CH_2CH_2CH_2$
       Cl Cl

IV.

V.

Answer: c   Section: 7

18. Which of the following statement(s) for dichlorocyclohexane is/are correct?

    I.   cis 1,2- is more stable than trans 1,2.
    II.   cis 1,3- is more stable than trans 1,3.
    III. cis 1,4- is more stable than trans 1,4.

    a. I
    b. II
    c. III
    d. I and II
    e. I and III

Answer: b   Section: 98

19. Which of the following is not an intermediate in the reaction of

    a. I
    b. II
    c. III
    d. IV
    e. V

and NBS ?

    I.   Br$_2$
    II. Br·
    III.

    IV.

    V.   HBr

    Answer: c Section: 5

SHORT ANSWER

20. Complete the following tree of reactions by supplying the missing reagents:

    Answer:
    a. NBS/heat
    b. HBr
    c. Br$_2$/CCl$_4$
    d. HBr/Peroxide
    e. Br$_2$/H$_2$O
    Section: 5

169

21. An alkane with the molecular formula C_5H_{10} forms only one monochlorinated product when heated with Cl_2/hv. Give the structure and the IUPAC name for this alkane.

Answer:

cyclopentane

Section: 2

22. How many monochlorinated products would be obtained from 2-methylbutane? Show the structures and give their IUPAC names.

Answer:

$$CH_3\underset{\underset{Cl}{|}}{\overset{\overset{CH_3}{|}}{C}}CH_2CH_3$$

2-chloro-2-methylbutane

$$ClCH_2\overset{\overset{CH_3}{|}}{CH}CH_2CH_3$$

1-chloro-2-methylbutane

$$CH_3\overset{\overset{CH_3}{|}}{CH}\underset{\underset{Cl}{|}}{CH}CH_3$$

2-chloro-3-methylbutane

$$CH_3\overset{\overset{CH_3}{|}}{CH}CH_2\overset{\overset{Cl}{|}}{CH_2}$$

1-chloro-3-methylbutane

Section: 2

23. Calculate the percent yield obtained for each product in problem 22. Assume that the relative ease of hydrogen abstraction in the chlorination process is 5 for 3° ; 3.8 for 2° ; and 1 for 1° hydrogens.

Answer:
2-chloro-2-methylbutane
$1 \cong 5 = 5$; %yield = $5 \cong 100/21.6 = 23.1\%$

1-chloro-2-methylbutane
$6 \cong 1 = 6$; %yield = $6 \cong 100/21.6 = 27.8\%$

2-chloro-3-methylbutane
$2 \cong 3.8 = 7.6$; %yield = $7.6 \cong 100/21.6 = 35.2\%$

1-chloro-3-methylbutane
$3 \cong 1 = 3$; %yield = $3 \cong 100/21.6 = 13.9\%$

Section: 3

24. Give the structure of the free radical intermediate for each product in a previous problem.

Answer:

$$\underset{\overset{|}{\text{Cl}}}{\overset{\overset{\text{CH}_3}{|}}{\text{CH}_3\text{CCH}_2\text{CH}_3}} \quad = \quad \overset{\overset{\text{CH}_3}{|}}{\text{CH}_3\text{CCH}_2\text{CH}_3}$$

$$\overset{\overset{\text{CH}_3}{|}}{\text{ClCH}_2\text{CHCH}_2\text{CH}_3} \quad = \quad \overset{\overset{\text{CH}_3}{|}}{\text{CH}_2\text{CHCH}_2\text{CH}_3}$$

$$\underset{\overset{|}{\text{Cl}}}{\overset{\overset{\text{CH}_3}{|}}{\text{CH}_3\text{CHCHCH}_3}} \quad = \quad \overset{\overset{\text{CH}_3}{|}}{\text{CH}_3\text{CHCHCH}_3}$$

$$\overset{\overset{\text{CH}_3}{|} \quad \overset{\text{Cl}}{|}}{\text{CH}_3\text{CHCH}_2\text{CH}_2} \quad = \quad \overset{\overset{\text{CH}_3}{|}}{\text{CH}_3\text{CHCH}_2\text{CH}_2}$$

Section: 3

25. Consider the following monobromination reaction, then answer the following questions.

$$CH_3CH_2CH_2CH_3 \; + \; \xrightarrow[\text{hv}]{\text{Br}_2} \; \text{Products}$$

a. Give the structures and the IUPAC names for the products.

b. Give the common names for the products.

c. Calculate $\Delta H°$ for the overall reaction using the following data:

$$CH_3CH_2\underset{|}{\overset{}{C}}HCH_3 \qquad \Delta H° = 100 \text{ kcal/mole}$$
$$H$$

$$CH_3CH_2CH_2\underset{|}{\overset{}{C}}H_2 \qquad \Delta H° = 105 \text{ kcal/mole}$$
$$H$$

$$Br\!\!-\!\!Br \qquad \Delta H° = 46 \text{ kcal/mole}$$

$$CH_3CH_2\underset{|}{\overset{}{C}}HCH_3 \qquad \Delta H° = 70 \text{ kcal/mole}$$
$$Br$$

$$CH_3CH_2CH_2CH_2Br \qquad \Delta H° = 74 \text{ kcal/mole}$$

$$H\!\!-\!\!Br \qquad \Delta H° = 87.5 \text{ kcal/mole}$$

d. Calculate the percent yield for each product. (relative rate of abstraction of 3° hydrogen is 1600; 2° is 82; and 1° is 1.)

e. Propose a step-by-step mechanism for the <u>major</u> product only.

f. Draw a schematic potential energy diagram for the rate-determining step (RDS) only.

g. Does the transition state of the RDS resemble more closely the reactants or the products?

h. Would the value of the activation energy be different for different alkanes? Explain.

i. Would the reaction slow down or speed up if I_2 is used instead of Br_2? Explain.

Answer:

a. $CH_3CHCH_2CH_3$ $CH_3CH_2CH_2CH_2$
    |
    Br   Br

    2-bromobutane   1-bromobutane

b. $CH_3CHCH_2CH_3$ $CH_3CH_2CH_2CH_2$
    |
    Br   Br

    sec-butyl bromide   ,   butyl bromide

c. $\Delta H°$ for 2-bromobutane formation = (100 + 46) - (70 + 87.5) = -11.5 kcal/mole

   $\Delta H°$ for 1-bromobutane formation = (105 + 46) - (74 + 87.5) = -10.5 kcal/mole

d. 2-bromobutane = 4 $\cong$ 82 = 328 %yield = 328 $\cong$ 100/334 = 98.2%

   1-bromobutane = 6 $\cong$ 1 = 6 %yield = 6 $\cong$ 100/334 = 1.8%

e. See graph below.

f. See graph below.

g. Product, since the transition state is closer to the products than it is to the reactants (Hammond Principle).

h. Yes, since each alkane has its own bond dissociation energy.

i. The reaction would slow down if I_2 is used. This is true because the activation energy for $I\cdot$ + $CH_3CH_2CH_2CH_3$ is larger than in $Br\cdot$ + $CH_3CH_2CH_2CH_3$.

e.

i.   Br⌒⌒Br $\xrightarrow{h\nu}$ 2 · Br

ii.   ·Br + CH₃CHCH₂CH₃ ⟶ HBr + CH₃ĊHCH₂CH₃

iii.   CH₃ĊHCH₂CH₃ + Br⌒⌒Br ⟶ CH₃CHCH₂CH₃ + ·Br

                                                Br

iv.   ·Br + ·Br ⟶ Br₂

(One of many possibilities)

f.

RDS

P. E. ↑

+12.5 kcal

Reaction Coordinate →

Section: 2

26. Write an equation to describe the initiation step in the chlorination of methane.

Answer: Cl-Cl + photon (hυ) $\xrightarrow{\hspace{3cm}}$ 2 Cl·   Section: 2

27. When light is shown on a mixture of chlorine and chloromethane, carbon tetrachloride is one of the components of the final reaction mixture. Propose a series of mechanistic steps which explain this observation.

Answer:
$CH_3Cl + Cl· \longrightarrow ·CH_2Cl + HCl$
$·CH_2Cl + Cl-Cl \longrightarrow CH_2Cl_2 + Cl·$
$CH_2Cl_2 + Cl· \longrightarrow ·CHCl_2 + HCl$
$·CHCl_2 + Cl-Cl \longrightarrow CHCl_3 + Cl·$
$CHCl_3 + Cl· \longrightarrow ·CCl_3 + HCl$
$·CCl_3 + Cl-Cl \longrightarrow CCl_4 + Cl·$
Section: 2

28. Species with unpaired electrons are called _____.

Answer: radicals <u>or</u> free radicals   Section: 2

29. Chlorination of methane can result in a mixture of chlorinated products. What experimental conditions should be used to favor the production of chloromethane over the other chlorinated products?

Answer: Make sure the molar ratio of methane to chlorine is relatively large.

Section: 2

30. Write a detailed, stepwise mechanism for the following reaction.

Answer:

Section: 2

MULTIPLE CHOICE

31. Consider the bond dissociation energies listed below in kcal/mol.

CH_3-Br 70
CH_3CH_2-Br 68
$(CH_3)_2CH$-Br 68
$(CH_3)_3C$-Br 65

These data show that the carbon-bromine bond is <u>weakest</u> when bromine is bound to a _____.

a. methyl carbon
b. primary carbon
c. secondary carbon
d. tertiary carbon
e. quanternary carbon

Answer: d Section: 2

32. Given the bond dissociation energies below (in kcal/mol), estimate the $\Delta H°$ for the propagation step $(CH_3)_2CH\cdot + Cl_2 \longrightarrow (CH_3)_2CHCl + Cl\cdot$.

$CH_3CH_2CH_2-H$ 98
$(CH_3)_2CH-H$ 95
$Cl-Cl$ 58
$H-Cl$ 103
$CH_3CH_2CH_2-Cl$ 81
$(CH_3)_2CH-Cl$ 80

   a. -22 kcal/mol
   b. +22 kcal/mol
   c. -40 kcal/mol
   d. +45 kcal/mol
   e. -45 kcal/mol

Answer: a   Section: 2

SHORT ANSWER

33. Given the bond dissociation energies below (in kcal/mol), calculate the overall $\Delta H°$ for the following reaction:

$(CH_3)_3CH + Br_2 \longrightarrow (CH_3)_3CBr + HBr$

$(CH_3)_3C-H$ 91
$(CH_3)_3C-Br$ 65
$Br-Br$ 46
$H-Br$ 88
CH_3-Br 70

Answer: -16 kcal/mol   Section: 2

34. Consider the reaction: $CH_3CH_2\cdot + Br_2 \longrightarrow Ch_3CH_2Br + Br\cdot$. Given that this reaction has an activation energy of +6 kcal/mol and a $\Delta H°$ of -22 kcal/mol, sketch a reaction-energy profile for this reaction. Label the axes and show E_a and $\Delta H°$ on your drawing.

Answer:

Section: 2

MULTIPLE CHOICE

35. The major monobominated product which results when methylcyclohexane is subjected to free radical bromination is:

a. a primary bromide
b. a secondary bromide
c. a tertiary bromide
d. a quaternary bromide
e. bromomethane

Answer: c Section: 4

36. Which of the halogens below undergoes free radical halogenation with ethane most rapidly?

a. fluorine
b. chlorine
c. iodine
d. bromine
e. pyridine

Answer: a Section: 4

37. Rank the free radicals (I–III) shown below in order of <u>decreasing</u> stability (ie, from most stable to least stable).

$\cdot CH_2CH_2CH(CH_3)_2$ I

$CH_3CH_2\overset{\bullet}{C}(CH_3)_2$ II

$CH_3\overset{\bullet}{C}HCH(CH_3)_2$ III

    a. I > III > II
    b. II > III > I
    c. I > II > III
    d. II > I > III
    e. III > II > I

Answer: b Section: 3

SHORT ANSWER

38. Write the structures of all of the monobromination products of 1,1,3,3-tetramethylcyclobutane.

Answer:

Section: 7

39. When 1,1,3,3-tetramethylcyclobutane is brominated at 125° C, the relative reactivity of the 1° :2° :3° hydrogens is approximately 1:82:1600. Estimate the amount of each monobromination product.

Answer: primary bromide, 3.5%; secondary bromide, 96.5% Section: 3

40. What is the relative reactivity of 2° vs 1° hydrogens in the free radical bromination of n-butane if the ratio of 1-bromobutane to 2-bromobutane formed is 7:93?

Answer: The 2° hydrogens are 20 times more reactive than the 1° ones.

Section: 3

41. What C_5H_{12} isomer will give only a single monochloration product?

Answer: $(CH_3)_4C$ <u>or</u> neopentane <u>or</u> 2, 2-dimethylpropane Section: 2

42. Use the Hammond Postulate to explain why free radical brominations are more selective than free radical chlorinations.

Answer:
The first propagation step in free radical bromination is endothermic while the analogous step in free radical chlorination is exothermic. From the Hammond Postulate, this means that the transition state for the bromination is product-like (ie, radical-like) while the transition state for the chlorination is reactant-like. The product-like transition state for bromination has the C-H bond nearly broken and a great deal of radical character on the carbon atom. The energy of this transition state reflects most of the energy difference of the radical products. This is not true in the chlorination case where the transition state possesses little radical character.
Section: 4

43. List the following radicals in order of increasing stability (ie, from least stable to most stable).

$(CH_3)_3C\cdot$, $CH_2{=}CHCH_2\cdot$, $CH_3CH_2\cdot$, $CH_3\cdot$, $(CH_3)_2CH\cdot$

Answer: $CH_3\cdot$ < $CH_3CH_2\cdot$ < $(CH_3)_2CH\cdot$ < $(CH_3)_3C\cdot$ < $CH_2{=}CHCH_2\cdot$ Section: 5

44. When $Br\cdot$ reacts with 1-butene ($CH_3CH_2CH{=}CH_2$), the hydrogen atom which is preferentially abstracted is the one which produces a resonance stabilized radical. Draw the major resonance contributing forms of this radical.

Answer:

$CH_3\overset{\bullet}{C}HCH{=}CH_2 \longleftrightarrow CH_3CH{=}CH\overset{\bullet}{C}H_2$

Section: 5

MULTIPLE CHOICE

45. During the free radical chlorination of methane, which of the following reactions has the lowest collision frequency?

a. $Cl\cdot + \cdot CH_3 \longrightarrow CH_3Cl$
b. $Cl\cdot + Cl_2 \longrightarrow Cl_2 + Cl\cdot$
c. $\cdot CH_3 + CH_4 \longrightarrow CH_4 + \cdot CH_3$
d. $\cdot CH_3 + Cl_2 \longrightarrow CH_3Cl + Cl\cdot$

Answer: a   Section: 2

46. The reaction $Br_2 + CH_3Br \longrightarrow CH_2Br_2 + HBr$ was carried out. Which of the following mechanism steps is both productive and relatively likely to occur?

    a. $Br\cdot + \cdot CH_2Br \longrightarrow CH_2Br_2$
    b. $Br\cdot + \cdot CH_3 \longrightarrow CH_3Br$
    c. $Br\cdot + Br_2 \longrightarrow Br_2 + Br\cdot$
    d. $Br\cdot + CH_3Br \longrightarrow HBr + \cdot CH_2Br$

    Answer: d Section: 2

47. The reation $Br_2 + CH_3Br \longrightarrow CH_2Br_2 + HBr$ was carried out. Which of the following mechanism steps is productive, but relatively unlikely to occur?

    a. $Br\cdot + CH_3Br \longrightarrow HBr + \cdot CH_2Br$
    b. $Br\cdot + \cdot CH_2Br \longrightarrow CH_2Br_2$
    c. $Br\cdot + Br_2 \longrightarrow Br_2 + Br\cdot$
    d. $Br\cdot + \cdot CH_3 \longrightarrow CH_3Br$

    Answer: b Section: 2

48. Which of the following reactions is a termination step in the free radical chlorination of methane?

    a. $Cl_2 + Cl\cdot \longrightarrow Cl\cdot + Cl_2$
    b. $Cl_2 \longrightarrow 2\ Cl\cdot$
    c. $\cdot CH_3 + Cl\cdot \longrightarrow CH_3Cl$
    d. $CH_4 + Cl\cdot \longrightarrow HCl + \cdot CH_3$

    Answer: c Section: 2

49. Which of the following is a chain propagation step in the free radical chlorination of methane?

    a. $CH_4 + Cl\cdot \longrightarrow \cdot CH_3 + HCl$
    b. $Cl_2 \longrightarrow 2\ Cl\cdot$
    c. $Cl\cdot + \cdot CH_3 \longrightarrow CH_3Cl$
    d. $\cdot CH_3 + CH_4 \longrightarrow CH_4 + \cdot CH_3$

    Answer: a Section: 2

50. Which of the following most nearly describes the geometry of the methyl radical?

    a. Trigonal pyramid, bond angle 109.5° .
    b. Trigonal pyramid, bond angle 120° .
    c. Trigonal planar, bond angle 109.5° .
    d. Trigonal planar, bond angle 120° .

    Answer: d Section: 6

51. How many electrons are contained in p orbital of the methyl radical?

    a. Zero
    b. One
    c. Two
    d. Three

    Answer: b Section: 6

SHORT ANSWER

52. Predict the major monobromination product in the following reaction.

    $(CH_3)_3CCH_2CH_3$ + Br_2 $\xrightarrow{\quad h\upsilon \quad}$

    Answer: $(CH_3)_3CCHBrCH_3$ Section: 4

53. The following molecule contains how many 1°, 2°, and 3° hydrogens?
    $(CH_3)_2CHCH_2CHCH_2CH_3$
    $\qquad\qquad\qquad |$
    $\qquad\qquad\quad CH_2CH_3$

    Answer: 12 1°, 6 2°, 2 3° Section: 3

54. Predict the major monobromination product in the following reaction.

    — $CH_2CH_2CH_3$ + Br_2 $\xrightarrow{h\nu}$

    Answer:

    Section: 5

MULTIPLE CHOICE

55. How many distinct dichlorination products can result when isobutane is subjected to free radical chlorination?

    a. 1
    b. 2
    c. 3
    d. 4
    e. 6

    Answer: c Section: 2

56. What is the <u>major</u> product of the following reaction?

    a. I
    b. II
    c. III
    d. IV
    e. V

I. $\overset{\underset{|}{Cl}}{CH_2}CH_2CH_2CH_2CH_2\overset{\underset{|}{Cl}}{CH_2}$

II.

III.

IV.

V.

Answer: b Section: 7

SHORT ANSWER

57. An unknown sample is suspected of being either ethane or isobutane. How would you distinguish between the two alkanes?

    Answer:
    A kinetic study of the rate of reaction with Br_2/hv between the two alkanes should reveal that the ease of abstraction of a 3° hydrogen is about a thousand times that of a 1° hydrogen.
    Section: 3

58. Rank the following fractions which result from the fractional distillation of petroleum in order of increasing boiling point: asphalt, gasoline, kerosene, natural gas, heating oil.

    Answer: natural gas < gasoline < kerosene < heating oil < asphalt

    Section: 98

MULTIPLE CHOICE

59. Which of the following statements concerning the use of hydrocarbons as fuels is <u>not</u> correct?

    a. The alkane n-octane has an octane number of 100 which indicates its high performance in automobile engines.
    b. Catalytic cracking converts straight-chain hydrocarbons that are poor fuels into branched-chain compounds.
    c. When poor fuels are used in an engine, combustion can occur before the spark plug fires which results in knocking.
    d. The quality of a fuel is indicated by its octane number, and a higher octane number typically results in less engine knocking.
    e. None of the above.

Answer: a

SHORT ANSWER

60. Explain why alkanes are generally considered unreactive compounds.

Answer:
They are unreactive because they have only strong σ bonds and atoms which have no partial charges.
Section: 1

MULTIPLE CHOICE

61. If cyclohexane reacts with excess Cl_2 at high temperature, how many distinct dichlorocyclohexane products are possible? Include all stereoisomers.

    a. 5
    b. 6
    c. 7
    d. 8
    e. 9

Answer: e   Section: 2

62. When butane undergoes free radical bromination, the product mixture contains 98% 2-bromobutane and 2% 1-bromobutane. How many times more susceptible to hydrogen atom abstraction is a secondary hydrogen in butane than is a primary hydrogen?

    a. 100
    b. 73.5
    c. 50
    d. 8.7
    e. 1.5

    Answer: b   Section: 3

SHORT ANSWER

63. Name the two brominated products which result when 2,3-dimethyl-2-butene reacts with NBS.

    Answer: 1-bromo-2,3-dimethyl-2-butene and 3-bromo-2,3-dimethyl-1-butene

    Section: 5

MULTIPLE CHOICE

64. What reagent can be used to convert cyclopentene to 3-bromocyclopentene in a single step?

    a. Br_2, hυ
    b. NBS, Δ
    c. HBr
    d. HBr with peroxide
    e. none of the above

    Answer: b   Section: 7

65. Why does cyclopropane undergo electrophilic addition reactions while cyclopentane does not?

    a. Cyclopropane contains considerable angle strain which is relieved in these reaction.
    b. Cyclopropane's planar geometry allows reagent to approach it more easily.
    c. Cyclopropane is more stable and therefore more reactive.
    d. These reactions result in a more favorable change in entropy in the case of cyclopropane.
    e. The cyclopropane molecules are smaller and therefore possess greater kinetic energy at any given temperature.

    Answer: a   Section: 7

SHORT ANSWER

66. What role do vitamins C and E play in biological radical reactions?

   Answer:
   Unwanted radicals in biological systems must be destroyed before they can cause damage to cells. Vitamins C and E are effective radical inhibitors since they form relatively stable radical upon reaction with more reactive radicals. Vitamin C traps radicals in the aqueous cellular environment, while vitamin E traps radicals formed in nonpolar membranes.

   Section: 8

67. Explain how CFC's are linked to depletion of stratospheric ozone.

   Answer:
   CFC's are persistent in the environment until they encounter ultraviolet light in the stratosphere. This interaction with a UV photon will cleave the C-Cl bond in the CFC and generate a chlorine radical. The chlorine radical reacts with ozone to form chlorine monoxide radical and molecular oxygen. The chlorine monoxide radical then reacts with a molecule of ozone to produce oxygen and to regenerate a chlorine radical. The two steps are the propagating steps in a radical chain reaction whereby a single chlorine radical can destroy about 100,000 ozone molecules.

   Section: 9

Chapter 9: Reactions at an sp3 Hybridized Carbon I

MULTIPLE CHOICE

1. Which of the following are the substitution products of the reaction shown below?

$$CH_3CH_2Br + {}^-OH \longrightarrow ?$$

  a. $CH_3CH_2Br^+H + O^-$
  b. $HOCH_2CH_2Br$
  c. $CH_3CH_2OH + Br^-$
  d. $CH_2{=}CH_2 + Br^- + H_2O$
  e. $CH_2{=}CHBr + H_2O$

Answer: c Section: 2

2. What is the nucleophile in the reaction shown below?

  a. I
  b. II
  c. III
  d. IV
  e. V

  I.

  II. H_2O
  III. $H_3O^{\oplus}$
  IV. $I^{\ominus}$
  V. $HO^{\ominus}$

Answer: b Section: 1

3. Which of the following is <u>not</u> a nucleophile?

  a. CH_3NH_2
  b. PH_3
  c. $^+CH_3$
  d. $CH_3O^{\ominus}$
  e. $HC{\equiv}C{:}^{\ominus}$

Answer: c Section: 1

4. Which of the following alkyl halides gives the <u>slowest</u> S_N2 reaction?

   a. CH_3CH_2Cl
   b. CH_3CHCH_2Cl
   $\quad\quad\quad\;\;|$
   $\quad\quad\quad\; CH_3$
   c. $\quad\quad Cl$
   $\quad\quad\quad\;|$
   $\quad\; CH_3CCH_2CH_3$
   $\quad\quad\quad\;\;|$
   $\quad\quad\quad\; CH_3$
   d. $\quad\quad\; Cl$
   $\quad\quad\quad\;|$
   $\quad\; CH_3CHCHCH_3$
   $\quad\quad\quad\;\;|$
   $\quad\quad\quad\; CH_3$
   e. $CH_3CHCH_2CH_3$
   $\quad\quad\quad\;|$
   $\quad\quad\quad CH_2$
   $\quad\quad\quad\;|$
   $\quad\quad\quad Cl$

   Answer: c Section: 2

5. Assuming no other changes, what is the effect of <u>doubling only</u> the concentration of the alkyl halide in the following S_N2 reaction?

   $CH_3Br + HO^{\ominus} \longrightarrow CH_3OH + Br^{\ominus}$

   a. No change
   b. Doubles the rate
   c. Triples the rate
   d. Quadruples the rate
   e. Rate is halved

   Answer: b Section: 2

6. Assuming no other changes, what is the effect of <u>doubling both</u> the alkyl halide and the nucleophile concentrations in the reaction given in the previous problem?

   a. No change
   b. Doubles the rate
   c. Triples the rate
   d. Quadruples the rate
   e. Rate is halved

   Answer: d Section: 2

7. Which of the following is the strongest nucleophile in an aqueous solution?

   a. HO^-
   b. F^-
   c. Cl^-
   d. Br^-
   e. I^-

   Answer: e   Section: 3

8. Which of the following is the best leaving group?

   a. HO^-
   b. F^-
   c. Cl^-
   d. Br^-
   e. I^-

   Answer: e   Section: 3

9. Which of the following species is most reactive in an S_N2 reaction?

   a. $CH_3CH_2{-}Cl$
   b. $CH_3CH_2{-}Br$
   c. $CH_3CH_2{-}I$
   d. $CH_3CH_2{-}F$
   e. $CH_3CH_2{-}OH$

   Answer: c   Section: 3

10. Which of the following S_N2 reactions is the <u>fastest</u>?

   a. $CH_3CHCH_3 + HO^- \longrightarrow CH_3CHCH_3 + Br^-$

         $|$ $|$

        Br   OH
   b. $CH_3CH_2CH_2I + HO^- \longrightarrow CH_3CH_2CH_2OH + I^-$
   c. $CH_3CHCH_3 + HO^- \longrightarrow CH_3CHCH_3 + I^-$

         $|$ $|$

        I   OH
   d. $CH_3CH_2CH_3Br + HO^- \longrightarrow CH_3CH_2CH_3OH + Br^-$
   e. $CH_3CH_2CH_2I + H_2O \longrightarrow CH_3CH_2CH_2OH + HI$

   Answer: b   Section: 2

11. Which of the following S_N2 reactions is the <u>slowest</u>?

    a. $CH_3CH_2CH_3Br$ + HO^- $\longrightarrow$ $CH_3CH_2CH_3OH$ + Br^-

    b. $CH_3CH_2CH_2Cl$ + HO^- $\longrightarrow$ $CH_3CH_2CH_2OH$ + Cl^-

    c. $CH_3CH_2CH_2I$ + HO^- $\longrightarrow$ $CH_3CH_2CH_2OH$ + I^-

    d. $CH_3CH_2CH_3F$ + HO^- $\longrightarrow$ $CH_3CH_2CH_2OH$ + F^-

    e. CH_3CHCH_3 + HO^- $\longrightarrow$ CH_3CHCH_3 + F^-
         | |
         F   OH

Answer: e Section: 2

12. When (S)-2-bromobutane undergoes an S_N2 reaction with CH_3O^- the product is the compound shown below. What is/are the configuration(s) of the product(s) obtained from this reaction?

$$CH_3CH_2CHCH_3$$
$$|$$
$$OCH_3$$

    a. S only
    b. R only
    c. A mixture of enantiomers with more R than S
    d. A mixture of enantiomers with more S than R
    e. Equal mixture of R and S

Answer: b Section: 7

13. The specific rotation of optically pure (R)-sec-butyl alcohol is -13.52°. An optically pure sample of (R)-sec-butyl bromide was converted into the corresponding sec-butyl alcohol via an S_N2 reaction. What is the specific rotation of the product, assuming 100% yield?

    a. -13.52°
    b. Between 0° and -13.52°
    c. Between 0° and +13.52°
    d. +13.52°
    e. zero

Answer: d Section: 7

14. Assuming no other changes, what is the effect of <u>doubling only</u> the concentration of the alkyl halide in the following S_N1 reaction?

$$(CH_3)_3CBr + I^{\ominus} \longrightarrow (CH_3)_3CI + Br^{\ominus}$$

a. No change
b. Doubles the rate
c. Triples the rate
d. Quadruples the rate
e. Rate is halved

Answer: b Section: 5

15. Assuming no other changes, what is the effect of <u>doubling only</u> the concentration of the nucleophile in the reaction given in the previous problem?

a. No change
b. Doubles the rate
c. Triples the rate
d. Quadruples the rate
e. Rate is halved

Answer: a Section: 5

16. Assuming no other changes, what is the effect of <u>doubling both</u> the concentration of the alkyl halide and the nucleophile in the reaction given in problem #15?

a. No change
b. Doubles the rate
c. Triples the rate
d. Quadruples the rate
e. Rate is halved

Answer: b Section: 5

17. Which of the following alkyl halides gives the fastest S_N1 reaction?

a. $CH_3CH_2CH_2Br$
b. CH_3CHCH_3
$\quad\quad |$
$\quad\quad Br$
c. CH_3CHCH_3
$\quad\quad |$
$\quad\quad I$
d. $CH_3CH_2CH_2I$
e. $CH_3CH_2CH_2Cl$

Answer: c Section: 5

18. Which of the following alkyl halides gives the slowest S_N1 reaction?

   a.   Br
              |
      $CH_3CHCHCH_3$
            |
           CH_3

   b. $CH_3CHCH_2CH_2Br$
          |
        CH_3

   c.   Br
           |
      $CH_3CCH_2CH_3$
         |
        CH_3

   d. $ClCH_2CHCH_2CH_3$
           |
         CH_3

   e. $CH_3CHCH_2CH_3$
        |
       CH_2Cl

   Answer: b Section: 5

19. Which of the following factors has <u>no</u> effect on the rate of S_N1 reactions?

   a. The nature of the alkyl halide
   b. The nature of the leaving group
   c. The concentration of the alkyl halide
   d. The concentration of the nucleophile
   e. The value of the rate constant

   Answer: d Section: 5

20. Which of the following carbocations is the most stable?

    a. $\overset{\oplus}{CH_3CH_2}$

    b. $CH_3\overset{\oplus}{C}HCH_3$

    c. CH_3
$$CH_3\overset{|}{\underset{|}{C\oplus}}$$
            CH_3

    d. CH_3
$$CH_3\overset{|}{\underset{|}{C}}\overset{\oplus}{C}H_2$$
            CH_3

    e. $\oplus$
      $CH_3CH_2CH_2$

Answer: c Section: 5

21. Which of the following carbocations is the least stable?

    a. I
    b. II
    c. III
    d. IV
    e. V

              $\oplus$
    I. CH_3CH_2

              $\oplus$
    II. CH_3CHCH_3

              CH_3
    III. $CH_3C\oplus$
              CH_3

              CH_3
              $|\oplus$
    IV. CH_3CCH_2
              CH_3

    V.   ⬡$\oplus$

Answer: a   Section: 5

22. Which of the following carbocations does not rearrange?

    a. $\overset{\oplus}{}$
       CH_3CH_2

    b. $\overset{\oplus}{}$
       CH_3CHCH_3

    c. CH_3
          |
      $CH_3C\oplus$
          |
         CH_3

             $\oplus$
    d. $CH_3CH_2CHCH_3$
    e. All the above

Answer: e Section: 6

23. Which of the following can't undergo nucleophilic substitution reactions?

    a. I
    b. II
    c. III
    d. IV
    e. V

    I. $CH_2=CHCH_2Br$

    II.   ⬡—Br (cyclohexenyl)

    III.   ⬡—CH_2Br (benzyl)

    IV.   ⬡—Br (cyclohexyl)

    V.   Br on cyclohexene

Answer: b Section: 8

24. Which of the following is the rate law for S_N1 mechanisms?

    a. Rate = k[Alkyl Halide] [Nucleophile]
    b. Rate = k[Nucleophile]
    c. Rate = k[Alkyl Halide]
    d. Rate = k[Alkyl Halide] [Nucleophile] + k_2[Alkyl Halide]
    e. Rate = k_1[Alkyl Halide] + k_2[Nucleophile]

Answer: c   Section: 5

25. Which of the following solvents is <u>aprotic</u>?

    a. CH_3CH_2OH
    b. $CH_3CH_2CH_2NH_2$
    c. $CH_3CH_2OCH_2CH_3$
    d. $CH_3CH_2NHCH_3$
    e. CH_3COH
        ‖
        O

Answer: c   Section: 10

26. Which of the following solvents is <u>protic</u>?

    a.   O
         ‖
     CH_3CH
    b.   O
         ‖
     CH_3CCH_3
    c. $CH_3CH_2OCH_3$
    d. CH_3CH_2OH
    e. $CH_3CH_2CH_2Cl$

Answer: d   Section: 10

27. Protic and aprotic solvents are very similar as solvents except for their

    a. polarity.
    b. dielectric constant.
    c. ability to stabilize anions by hydrogen bonding.
    d. ability to stabilize cations by hydrogen bonding.
    e. ability to stabilize cations with unshared pairs of electrons.

Answer: c   Section: 10

28. Which of the following best explains why S_N1 reactions involving a neutral reactant are <u>faster</u> in polar solvents?

    a. The substrate is more soluble in polar solvents.
    b. The substrate is less soluble in polar solvents.
    c. The nucleophile is solvated by polar solvents.
    d. Solvation by polar solvents stabilizes the carbocation.
    e. Solvation by polar solvents stabilizes the transition state.

    Answer: e Section: 10

29. Which of the following statements is generally true for S_N1 reactions?

    a. Complete inversion of configuration occurs.
    b. These are favored by nonpolar solvents.
    c. These reactions are favored by polar solvents.
    d. Reaction rates depend only on the concentration of the nucleophile.
    e. The mechanism is a one-step back attack.

    Answer: c Section: 10

SHORT ANSWER

30. Explain why allyl chloride undergoes S_N1 reactions even though it is a 1° halide?

$$CH_2=CHCH_2Cl$$
allyl chloride

Answer:
Allyl chloride forms a relatively stable allylic carbocation with two contributing resonance structures.

Section: 8

31. Consider the following reaction mechanism:

What effect will the following have on the rate?
a. addition of Br⁻
b. using a more polar solvent
c. using a stronger nucleophile
d. using a more concentrated nucleophile

$$I. \quad R\!-\!Br \rightleftharpoons \overset{\oplus}{R} + \overset{\ominus}{Br}$$
$$II. \quad \overset{\oplus}{R} + \overset{\ominus}{Nu} \longrightarrow R\!-\!Nu$$

Answer:
a. reaction will slow down because of competition with Nu⁻
b. reaction will speed up because a more polar solvent assists in ionization and the formation of the carbocation
c. no effect on S_N1 reactions
d. no effect on S_N1 reactions
Section: 10

32. Explain why S_N2 reactions proceed faster in the solvent dimethylsulfoxide than in ethanol?

$$CH_3SOCH_3 \qquad CH_3CH_2OH$$

dimethlysolfoxide   ethanol

Answer:
Dimethlysulfoxide is an aprotic solvent. Ethanol is a protic solvent which can hydrogen bond to the nucleophile, decreasing its nucleophilicity. Therefore, S_N2 reactions favor aprotic solvents.
Section: 10

MULTIPLE CHOICE

33. Which of the following best describes the carbon-chlorine bond of an alkyl chloride?

a. nonpolar; no dipole
b. polar; δ+ at carbon and δ- at chlorine
c. polar; δ- at carbon and δ+ at chlorine
d. ionic
e. none of the above

Answer: b Section: 1

34. What is the leaving group in the reaction shown below?

    a. I
    b. II
    c. III
    d. IV
    e. V

$$CH_3-\bigcirc-SO_3CH_3 + \bigcirc-O^{\ominus} \longrightarrow CH_3-\bigcirc-SO_3^{\ominus} + \bigcirc-OCH_3$$

    I. $CH_3-\bigcirc-SO_3^{\ominus}$,

    II. $\bigcirc-OCH_3$

    III. $\bigcirc-O^{\ominus}$

    IV. $CH_3-\bigcirc-SO_3CH_3$

    V. $CH_3O^{\ominus}$

Answer: a Section: 98

35. Which of the following is classified as a vinyl halide?

    a. $CH_3CH{=}CHOH$
    b. $CH_3CH{=}CHCl$
    c. $CH_3CH{=}CHCH_2Cl$
    d. $CH_3CH_2CH_2CH_2Br$
    e. $BrCH_2CH{=}CH_2$

Answer: b Section: 8

36. Which of the following species is the <u>least</u> nucleophilic?

    a. $(CH_3)_3CO^-$
    b. H_2O
    c. $(CH_3)_3N$
    d. BF_3
    e. CN^-

Answer: d Section: 3

SHORT ANSWER

37. Provide a detailed, stepwise mechanism for the reaction below.

    $(CH_3)_2CHCH_2CH_2CH_2I + CN^- \longrightarrow (CH_3)_2CHCH_2CH_2CH_2CN + I^-$

    Answer:

    $(CH_3) CHCH_2CH_2CH_2 - I + CN^- \longrightarrow (CH_3)_2CHCH_2CH_2CH_2 - CN$

    Section: 2

38. Do all primary alkyl iodides undergo S_N2 reactions with sodium cyanide in DMSO at identical rates? Explain.

    Answer:
    No. All primary iodides are not equally accessible to attack by the CN^-. Steric hindrance varies among primary iodides.
    Section: 2

39. Rank the species below in order of increasing nucleophilicity in protic solvents: $CH_3CO_2^-$, CH_3S^-, HO^-, H_2O.

    Answer:

    $H_2O < CH_3CO_2^- < HO^- < CH_3S^-$
    Section: 3

40. What type of solvent is best for S_N2 reactions which employ anionic nucleophiles: polar, protic solvents; polar, aprotic solvents; or nonpolar solvents? Explain.

    Answer:
    Polar, aprotic solvents are best. These solvents have strong dipole moments to enhance solubility of the anionic species but lack the ability to solvate the anion by hydrogen bonding.
    Section: 10

41. Provide the structure of the major organic product which results when (S)-2-iodopentane is treated with KCN in DMF.

Answer:

Section: 2

MULTIPLE CHOICE

42. Which of the following compounds will undergo an S_N2 reaction most readily?

   a. $(CH_3)_3CCH_2I$
   b. $(CH_3)_3CCl$
   c. $(CH_3)_2CHI$
   d. $(CH_3)_2CHCH_2CH_2CH_2Cl$
   e. $(CH_3)_2CHCH_2CH_2CH_2I$

Answer: e Section: 2

SHORT ANSWER

43. Provide the structure of the major organic product in the following reaction.

Answer:

Section: 2

44. Provide the structure of the major organic product in the following reaction.

$(CH_3)_3N + CH_3CH_2CH_2I \longrightarrow$

Answer: $[(CH_3)_3NCH_2CH_2CH_3]^+$ I^- Section: 2

45. Provide the major organic product of the reaction below and a detailed, stepwise mechanism which accounts for its formation.

Answer:

Section: 5

MULTIPLE CHOICE

46. S_N1 reactions usually proceed with:

    a. equal amounts of inversion and retention at the center undergoing substitution.
    b. slightly more inversion than retention at the center undergoing substitution.
    c. slightly more retention than inversion at the center undergoing substitution.
    d. complete inversion at the center undergoing substitution.
    e. complete retention at the center undergoing substitution.

Answer: b Section: 7

47. Which of the compounds below undergoes solvolysis in aqueous ethanol most rapidly?

    a. cyclohexyl bromide
    b. methyl iodide
    c. isopropyl chloride
    d. 3-chloropentane
    e. 3-iodo-3-methylpentane

    Answer: e Section: 5

SHORT ANSWER

48. Provide the structure of the major organic products which result in the reaction below.

    Answer:

    Section: 5

49. Provide the structure of the major organic product which results in the following reaction.

    Answer:

    Section: 6

50. List the following compounds in order of increasing reactivity in an S_N1 reaction.

    CH_3Br, $CH_3CH_2CH_2I$, $(CH_3)_3CI$, $CH_3CHBrCH_3$, CH_3CHICH_3

    Answer: $CH_3Br < CH_3CH_2CH_2I < CH_3CHBrCH_3 < CH_3CHICH_3 < (CH_3)_3CI$

    Section: 6

MULTIPLE CHOICE

51. Which of the following is a <u>secondary</u> alkyl halide?

    a. CH_3Br
    b. $(CH_3)_3CBr$
    c. $(CH_3)_2CHBr$
    d. $(CH_3)_2CHCH_2Br$

    Answer: c Section: 2

52. Which of the following is <u>not</u> normally considered to be a nucleophile?

    a. NH_3
    b. NH_2CH_3
    c. $HC{\equiv}C{:}^-$
    d. $CH_3CH_2{}^+$

    Answer: d Section: 1

53. Which of the following S_N2 reactions is the <u>fastest</u>?

    a. $CH_3CH_2CH_2CH_2Br + OH^- \longrightarrow CH_3CH_2CH_2CH_2OH + Br^-$
    b. $CH_3CH_2CH_2CH_2Br + H_2O \longrightarrow CH_3CH_2CH_2CH_2OH + HBr$
    c. $CH_3CH_2CHBrCH_3 + OH^- \longrightarrow CH_3CH_2CHOHCH_3 + Br^-$
    d. $CH_3CH_2CHBrCH_3 + H_2O \longrightarrow CH_3CH_2CHOHCH_3 + HBr$

    Answer: a Section: 2

54. Which of the following S_N2 reactions is the <u>slowest</u>?

    a. $CH_3CH_2CHBrCH_3 + OH^- \longrightarrow CH_3CH_2CHOHCH_3 + Br^-$
    b. $CH_3CH_2CHBrCH_3 + H_2O \longrightarrow CH_3CH_2CHOHCH_3 + HBr$
    c. $CH_3CH_2CH_2CH_2Br + OH^- \longrightarrow CH_3CH_2CH_2CH_2OH + Br^-$
    d. $CH_3CH_2CH_2CH_2Br + H_2O \longrightarrow CH_3CH_2CH_2CH_2OH + HBr$

    Answer: b Section: 2

55. The specific rotation of optically pure (R)-$C_6H_5CHOHCH_3$ is -42.3°. An optically pure sample of (R)-$C_6H_5CHClCH_3$ was converted into the corresponding alcohol via an S_N2 reaction. What is the specific rotation of the product?

    a. -42.3°
    b. Between 0° and -42.3°
    c. Between +42.3° and 0°
    d. +42.3°

    Answer: d Section: 7

56. When (S)-1-bromo-1 phenylethane undergoes an S_N1 reaction with methanethiol (CH_3SH), the product is the compound shown. What is/are the configuration(s) of the product as obtained from this reaction?

$C_6H_5CH-CH_3$
  |
  SCH_3

    a. S only.
    b. R only.
    c. A mixture of the enantiomers, with slightly more S than R.
    d. A mixture of the enantiomers, with slightly more R than S.

Answer: d Section: 7

57. Which of the following does <u>not</u> provide evidence that there are two different mechanisms for nucleophilic substitution?

    a. Reaction products when CH_3I is used as the substrate.
    b. Reaction products when $(CH_3)_3CCH_2I$ is used as substrate.
    c. The stereochemistry of nucleophilic substitutions.
    d. The effect of nucleophile concentration on rate.

Answer: a Section: 2

58. Which of the following is the best nucleophile in water?

    a. I^-
    b. CH_3SCH_3
    c. CH_3OCH_3
    d. Cl^-

Answer: a Section: 3

SHORT ANSWER

59. In each of the pairs below, which is the best nucleophile in alcoholic solvents?

    a. CH_3S^- or CH_3O^-
    b. $(CH_3)_2NH$ or $(CH_3)_3N$
    c. Cl^- or F^-
    d. SCN^- or OCN^-

Answer:
    a. CH_3S^-
    b. $(CH_3)_2NH$
    c. Cl^-
    d. SCN^-

Section: 3

60. Draw and describe the transition state in the S_N2 reaction between CH_3I and $CH_3CH_2O^-Na^+$.

Answer:
In this transition state the oxygen-carbon bond is partially formed and the carbon-iodine bond is partially broken. There is essentially a trigonyl bipyramidal electron geometry around the carbon being attacked by the ethoxide, and the O-C-I angle is 180°. Both the oxygen and iodine atoms possess a partial negative charge.

$$\delta-$$
$$I$$
$$H-C{}^{H}_{H}$$
$$OCH_2CH_3$$
$$\delta-$$

Section: 2

MULTIPLE CHOICE

61. Which of the following statements concerning SN2 reactions of alkyl halides is <u>not</u> correct?

a. The rate of reaction depends on the concentration of the nucleophile.
b. The rate of reaction depends on the concentration of the alkyl halide.
c. The rate of reaction of a particular alkyl bromide depends on the steric accessibility of the carbon of the C-Br bond.
d. All alkyl iodides react more rapidly than all alkyl chlorides.
e. The rate of reaction depends on the relative nucleophilicity of the nucleophile.

Answer: d   Section: 3

SHORT ANSWER

62. In an S_N2 reaction why does the nucleophile attack the carbon on the side opposite the leaving group?

Answer:
Back side attack occurs because the orbital of the nucleophile that contains its nonbonding electrons interacts with the empty σ* MO associated with the carbon-halogen bond. This orbital has its larger lobe on the side of the carbon directed away from the carbon-halogen bond, thus the nucleophile can achieve the best overlap with this lobe.
Section: 2

MULTIPLE CHOICE

63. What product results from the S_N2 reaction between (R)-2-chloropentane and hydroxide?

    a. (R)-2-pentanol
    b. (S)-2-pentanol
    c. racemic pentanol
    d. 1-pentanol
    e. 3-pentanol

    Answer: b   Section: 2

SHORT ANSWER

64. Why, in a polar protic solvent, is iodide a better nucleophile than fluoride?

    Answer:
    In addition to iodide being more polarizable due to its larger size, the fluoride ion is more highly solvated which retards its ability to function as a nucleophile.
    Section: 3

MULTIPLE CHOICE

65. Which of the following correctly describes the relative nucleophilicities of methoxide and tert-butoxide?

    a. These alkoxides have essentially the same nucleophilicities since the negative charge in both is localized on an oxygen atom.
    b. Methoxide is more nucleophilic because the nucleophilicity of tert-butoxide is diminished by steric effects.
    c. tert-Butoxide is more nucleophilic because it contains three methyl groups which increase the charge on its oxygen by donating electron density.
    d. tert-Butoxide is more nucleophilic because it is more basic.
    e. None of the above.

    Answer: b   Section: 3

SHORT ANSWER

66. A student attempted to prepare 1-chlorobutane by treating 1-butanol with NaCl in acetone. Was the student successful? Explain.

    Answer:
    The student was unsuccessful. The reaction would require chloride to displace hydroxide. Since hydroxide is a much stronger base, and therefore a much poorer leaving group, the reaction does not occur.
    Section: 4

MULTIPLE CHOICE

67. The hydrolysis of <u>tert</u>-butyl chloride proceeds more rapidly in a solvent mixture which is 70% water/30% acetone than in one which is 30% water/70% acetone. Why?

    a. The transition state in the carbocation formation step is better stabilized in the more polar solvent mixture.
    b. The reaction proceeds by an S_N2 mechanism wherein the rate is increased by increasing the concentration of the nucleophile water.
    c. The reaction proceeds by an S_N1 mechanism wherein the rate is increased by increasing the concentration of the nucleophile water.
    d. The solvent which contains a greater percentage of water is less polar, and this destabilizes the <u>tert</u>-butyl chloride.
    e. None of the above.

    Answer: a   Section: 10

68. Which of the following alkyl bromides undergoes solvolysis in methanol without rearrangement?

    a. (R)-2-bromo-3-ethylpentane
    b. (S)-2-bromo-3-ethylpentane
    c. (R)-3-bromo-2-methylpentane
    d. (S)-3-bromo-2-methylpentane
    e. 3-bromo-3-ethylpentane

    Answer: e   Section: 6

69. Which of the following alkyl bromides is likely to undergo rearrangement by a 1,2-methyl shift?

    a. benzyl bromide
    b. 2-bromo-3-ethylpentane
    c. 3-bromo-3-methylpentane
    d. 2-bromo-3,3-dimethylpentane
    e. 3-bromo-2,3-dimethylpentane

    Answer: d   Section: 6

SHORT ANSWER

70. Why are vinylic halides unreactive in both S_N2 and S_N1 reactions?

    Answer:
    Vinylic halides do not undergo S_N2 reactions because as the nucleophile approaches the back side of the *sp2* carbon it is repelled by the π electron cloud of the double bond. They do not undergo S_N1 reactions because vinylic carbocations are prohibitively unstable (less stable than primary carbocations).
    Section: 8

Chapter 10:   Reactions at an sp3 Hybridized Carbon II

MULTIPLE CHOICE

1. Which of the following are the elimination products of the reaction shown below?

$$CH_3CH_2Br + {}^-OH \longrightarrow ?$$

   a. $CH_3CH_2Br^+H + O^-$
   b. $HOCH_2CH_2Br$
   c. $CH_3CH_2OH + Br^-$
   d. $CH_2{=}CH_2 + Br^- + H_2O$
   e. $CH_2{=}CHBr + H_2O$

   Answer: d   Section: 1

2. Which of the following compounds undergoes E2 reactions with the <u>fastest</u> rate?

   a. CH_3CHCH_3
      |
      Cl
   b. $CH_3CH_2CH_2Cl$
   c. $CH_3CH_2CH_2I$
   d. CH_3CHCH_3
      |
      I
   e. $CH_3CH_2CH_2Br$

   Answer: d   Section: 2

3. Which of the following bases gives the highest anti-Zaitsev product in E2 reactions when reacted with 2-bromo-2,3-dimethylbutane?

   a. $CH_3O^{\ominus}$
   b. $CH_3CH_2O^{\ominus}$
   c. $CH_3CHO^{\ominus}$
      |
      CH_3
   d. CH_3
         |
      $CH_3CO^{\ominus}$
         |
         CH_3
   e. CH_3
            |
      $CH_3CH_2CO^{\ominus}$
            |
            CH_3

   Answer: e   Section: 2

4. Which of the following alkyl halides undergoes E1 reactions with the <u>fastest</u> rate?

a. CH_3CHCH_3
        |
        F

b. CH_3CHCH_3
        |
        Cl

c. CH_3CHCH_3
        |
        Br

d. CH_3CHCH_3
        |
        I

e. CH_3I

Answer: d Section: 3

5. Which of the following alkyl halides forms the most stable carbocation when it undergoes an E1 reaction?

a. I
b. II
c. III
d. IV
e. V

I.

II.

III.

IV.

V.

Answer: d Section: 3

6. What is the major product of the following E2 reaction?

$$CH_3CHCH_2CH_3 + HO^- \xrightarrow{\text{heat}} Product(s)$$
$$|$$
$$Br$$

a.

H H
| /
$CH_3CH_2C=C$
\
H

b. $CH_3CH_2CH=CH_2$

c. CH₃
CH_3 H
\ /
C=C
/ \
H CH_3

d. CH₃
CH_3 CH_3
\ /
C=C
/ \
H   H

e. $CH_3C=CH_2$
|
CH_3

Answer: c Section: 2

7. Consider the following experimental data for the rate of the reaction given below:

$$(CH_3)_3CI + CH_3OH \xrightarrow{\text{heat}} (CH_3)_2C=CH_2 + \overset{+}{C}H_3OH_2 + I^-$$

Experiment #1	[Alkyl Halide]	[Base]	Rate
1	0.01	0.01	1
2	0.02	0.01	2
3	0.01	0.02	1

What is the mechanism for the reaction?

a. first order, S_N1

b. first order, S_N2

c. first order, E1
d. first order, E2
e. none of the above

Answer: c Section: 3

209

8. Which of the compounds shown below is/are the product(s) of this reaction:

a. I. only
b. II. only
c. I. and II. are of equal yield
d. I. is major, II. is minor
e. I. is minor, II. is major

Answer: a Section: 6

9. Which of the following compounds is/are the products of this reaction:

a. I. only
b. II. only
c. I. and II. are of equal yield
d. I. is major, II. is minor
e. I. is minor, II. is major

Answer: e Section: 6

10. What is the major product of the following reaction?

$$CH_3CHCH_2CH_3 + CH_3CH_2O \xrightarrow[heat]{\oplus \text{ alcohol}} ?$$
$$|$$
$$Br$$

a. $CH_2=CHCH_2CH_3$
b. $CH_3CH=CHCH_3$
c. $CH_3CHCH_2CH_3$
   $|$
   OCH_2CH_3

d. $CH_3CH_2CH_2CH_2OCH_2CH_3$

e. $CH_3C=CH_2$
   $|$
   CH_3

Answer: b Section: 8

11. What is the major product of the following reaction?

$$CH_3CHCH_2CH_3 + CH_3CH_2OH \xrightarrow[temp.]{room} ?$$
$$|$$
$$Br$$

a. $CH_2=CHCH_2CH_3$
b. $CH_3CH=CHCH_3$
c. $CH_3CHCH_2CH_3$
   $|$
   OCH_2CH_3

d. $CH_3CH_2CH_2CH_2OCH_2CH_3$

e. $CH_3C=CH_2$
   $|$
   CH_3

Answer: c Section: 8

12. What is the major product of the following reaction?

    a. I
    b. II
    c. III
    d. IV
    e. V

$$CH_2-\underset{\underset{Br}{|}}{\overset{\overset{CH_3}{|}}{C}}-CH_3 \ + \ KOH \ \xrightarrow[\text{heat}]{\text{alcohol}} \ ?$$

I. $CH_2-\underset{\underset{OH}{|}}{\overset{\overset{CH_3}{|}}{C}}-CH_3$

II. $CH_2-\overset{\overset{CH_3}{|}}{C}=CH_2$

III. $CH_2-\overset{\overset{CH_3}{|}}{CH}-CH_2OH$

IV. $\underset{\underset{OH}{|}}{CH}-\underset{\underset{H}{|}}{\overset{\overset{CH_3}{|}}{C}}-CH_3$

V. $CH=\overset{\overset{CH_3}{|}}{C}-CH_3$

Answer: e   Section: 8

13. What is/are the product(s) of the following reaction?

$$CH_3CH_2\overset{\ominus}{O} \ + \ CH_3\underset{\underset{CH_3}{|}}{\overset{\overset{CH_3}{|}}{C}}Br \ \longrightarrow \ ?$$

    a. $CH_3CH_2O\underset{\underset{CH_3}{|}}{\overset{\overset{CH_3}{|}}{C}}-CH_3$

    b. $CH_2{=}CH_2$
    c. $CH_3\underset{\underset{CH_3}{|}}{C}{=}CH_2$

    d. a. and b.
    e. a. and c.

Answer: c   Section: 8

14. What is/are the product(s) of the following reaction?

$$CH_3-\underset{\underset{CH_3}{|}}{\overset{\overset{CH_3}{|}}{C}}-O^{\ominus} + CH_3CH_2Br \longrightarrow ?$$

a. $CH_3CH_2O\underset{\underset{CH_3}{|}}{\overset{\overset{CH_3}{|}}{C}}-CH_3$

b. $CH_2{=}CH_2$

c. $CH_3\underset{\underset{CH_3}{|}}{C}{=}CH_2$

d. a. and b.

e. a. and c.

Answer: d Section: 8

15. Starting with 2-butene, which of the following is the best method for preparing 2-butyne?

a. HBr; H_2/Ni; Zn/H$^+$
b. HBr; Zn/H$^+$; H_2/Ni
c. Br_2/CCl$_4$; Zn/H$^+$; H_2/Ni
d. Br_2/CCl$_4$; 2NaNH$_2$
e. HBr; NaNH$_2$

Answer: d Section: 9

16. What is the major product of the following reaction?

$$CH_2{=}CHCH_2\underset{\underset{Br}{|}}{CH}CH_3 + KOH \xrightarrow[\text{heat}]{\text{alcohol}} ?$$

a. $CH_2 = CHCH_2CH = CH_2$

b. $CH_3CH_2CH_2\underset{\underset{OH}{|}}{CH}CH_3$

c. $CH_2 = CHCH = CHCH_3$

d. $CH_2 = CHCH\underset{\underset{OH}{|}}{CH_2}CH_3$

e. $CH_2 = CHCH_2CH = CH_2$

Answer: c Section: 10

SHORT ANSWER

17. Which of the following alkyl chlorides would undergo substitution most rapidly when treated with NaCCH: chloroethane, 2-chloropropane, or 1-chloro-2,2-dimethylpropane? Provide the structure of the substitution product.

    Answer: $CH_3CH_2C{\equiv}CH$ Section: 9

18. Provide a series of synthetic steps by which $(CH_3)_2C{=}CH_2$ could be prepared from 2-methylpropane.

    Answer:
    1. Br_2, $h\upsilon$
    2. $NaOCH_3$, CH_3OH

    Section: 12

19. When 1-bromo-2,2-dimethylcyclopentane is heated in ethanol, one of the products which results is shown below. Provide a detailed, stepwise mechanism for the production of this compound, and give the name of the mechanism by which it is produced.

    Answer:

    E1 mechanism.

    Section: 3

20. Provide the structure of the major organic product which results in the following reaction.

NaOCH$_3$, CH$_3$OH
heat

Answer:

Section: 6

21. Provide the structure of the major organic product which results in the following reaction.

NaOCH$_3$, CH$_3$OH
heat

Answer:

Section: 6

22. Provide the structure of the major organic product which results in the following reaction.

NaOCH$_3$, CH$_3$OH
heat

Answer:

Section: 6

23. Which diastereomer of 1-bromo-4-t-butylcyclohexane, the cis or the trans, undergoes elimination more rapidly when treated with sodium ethoxide? Explain your answer.

Answer:
The cis isomer reacts more quickly. An axial orientation of the C-Br bond is required for E2 in six-membered ring systems. In order for this to occur in the trans isomer, the bulky t-butyl group must also assume an axial orientation; this requires substantial energy of activation.
Section: 6

24. When 1-iodo-1-methylcyclohexane is treated with NaOCH$_2$CH$_3$, the more highly substituted alkene product predominates. When KOC(CH$_3$)$_3$ is used instead, the less highly substituted alkene product predominates. Offer an explanation.

Answer:
The unhindered ethoxide produces the more stable alkene product (ie, the more highly substituted alkene possible). When the bulky t-butoxide is used, the most accessible hydrogen is removed. This results in the least highly substituted alkene possible.
Section: 2

MULTIPLE CHOICE

25. Which of the following statements correctly describe(s) E1 reactions of alkyl halides (RX)?

    I. Rate = k[base]
   II. Rate = k[base][RX]
  III. Rate = k[RX]
  IV. The reactions occur in two distinct steps.
   V. Rearrangements are sometimes seen.

  a. II and IV
  b. III and V
  c. I, IV, and V
  d. I only
  e. III, IV, and V

Answer: e Section: 3

SHORT ANSWER

26. Provide the structure of the major organic product which results when 2-bromo-2-methylbutane is treated with sodium ethoxide.

Answer:

```
   H        CH3
    \      /
     C=C
    /      \
 CH3        CH3
```
Section: 2

MULTIPLE CHOICE

27. How many distinct alkene products are possible when the alkyl iodide below undergoes E2 elimination?

  a. 1
  b. 2
  c. 3
  d. 4
  e. 5

Answer: e Section: 2

SHORT ANSWER

28. Provide the structure of the major organic product in the following reaction.

$$
\underset{\text{Ph}}{\overset{\text{H}\quad\text{CH}_3}{\underset{\text{Br}}{\bigwedge}}}\text{Ph} \quad \xrightarrow[\text{CH}_3\text{OH}]{\text{NaOCH}_3}
$$

Answer:

$$
\underset{\text{CH}_3}{\overset{\text{Ph}}{\diagdown}}=\underset{\text{Ph}}{\overset{\text{H}}{\diagup}}
$$

Section: 4

29. Show the best way to prepare $CH_3OCH(CH_3)_2$ by an S_N2 reaction.

Answer: CH_3I + $NaOCH(CH_3)_2$ Section: 9

MULTIPLE CHOICE

30. Predict the two most likely mechanisms for the reaction of 2-iodohexane with sodium ethoxide.

a. S_N2 and S_N1
b. E1 and E2
c. S_N2 and E2
d. E1 and S_N1
e. E2 and S_N1

Answer: c Section: 8

31. Predict the two most likely mechanisms which occur when 2-iodohexane is heated in ethanol.

a. S_N2 and S_N1
b. E1 and E2
c. S_N2 and E2
d. E1 and S_N1
e. E2 and S_N1

Answer: d Section: 8

32. Which of the following is the best method for preparing THF,

    a. I
    b. II
    c. III
    d. IV
    e. V

 ?

I. $CH_3CH_2\overset{\ominus}{O} + CH_3CH_2Br$

II. $CH_3CH_2CH_2\overset{\ominus}{O} + CH_3Br$

III. $BrCH_2CH_2CH_2CH_2\overset{\ominus}{O}$

IV. $CH_3CHCH_2CH_2CH_2Br$
    $\quad\quad |\overset{\ominus}{O}$

V. $CH_3CH_2CH_2Br + CH_3\overset{\ominus}{O}$

Answer: c Section: 11

SHORT ANSWER

33. Draw all likely alkene products in the following reaction and circle the product you expect to predominate.

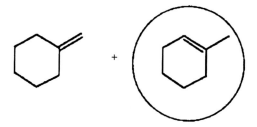

Answer:

Section: 2

34. Draw all likely alkene products in the following reaction and circle the product you expect to predominate.

Answer:

Section: 2

35. Draw all likely alkene products in the following reaction and circle the product you expect to predominate.

Answer:

Section: 2

36. Draw the alkene product which results when 1-bromopentane is heated in acetone containing NaOH. Give a detailed, step-by-step mechanism for the production of this compound.

Answer:

Section: 1

MULTIPLE CHOICE

37. Which of the alkyl chlorides listed below undergoes dehydrohalogenation in the presence of a strong base to give 2-pentene as the <u>only</u> alkene product?

    a. 1-chloropentane
    b. 2-chloropentane
    c. 3-chloropentane
    d. 1-chloro-2-methylbutane
    e. 1-chloro-3-methylbutane

Answer: c Section: 2

38. Dehydrohalogenation of 2-bromobutane in the presence of a strong base proceeds via which of the following mechanistic pathways?

    a. S_N1
    b. S_N2
    c. E1
    d. E2
    e. none of the above

Answer: d Section: 4

SHORT ANSWER

39. Which base, ammonia (NH_3) or triethylamine [$(CH_3CH_2)_3N$], would be a better choice for use in converting 1-chlorohexane to 1-hexene? Explain briefly.

    Answer:
    Triethylamine. Amines can serve as both nucleophiles and as bases in reactions with alkyl halides. Increasing the steric bulk about the nitrogen diminishes the nucleophilicity while allowing the amine to continue to function effectively as a base.
    Section: 8

40. Provide the reagents necessary for carrying out the transformation of cyclopentane to cyclopentene.

    Answer:
    1. Br_2, hυ
    2. NaOH, acetone
    Section: 12

41. Provide the reagents necessary for carrying out the transformation of 2-methylheptane to 2-methyl-1-heptene.

    Answer:
    1. Br_2, hυ
    2. $(CH_3)_3CO^-K^+$, $(CH_3)_3COH$
    Section: 12

42. Provide the reagents necessary for carrying out the transformation of 2-methylheptane to 2-methyl-2-heptene.

    Answer:
    1. Br_2, hυ
    2. NaOH, acetone
    Section: 12

43. Write the structures for the products of the following S_N2 reactions.

    a) NaC≡CH + ⬠— CH_2Br ⟶

    b) $(CH_3)_2CHCH_2CHCH_3$ + NaI ⟶
                          |
                          Cl

    c) $CH_2CH_2CH_2I$ + $NaOCH_3$

    Answer:
    a) 3-cyclopentyl-1-propyne
    b) 2-iodo-4-methylpentane
    c) 1-methoxypropane
    Section: 9

44. List the following compounds in order of increasing reactivity in an El elimination.

$CH_3CH_2CHBrCH_3$, $CH_3CH_2CH_2CH_2Br$, $(CH_3)_3CBr$

Answer: $CH_3(CH_2)_3Br < CH_3CH_2CHBrCH_3 < (CH_3)_3CBr$ Section: 3

45. Supply the missing alkyl halide reactant in the elimination reactions shown below.

a) ? $\xrightarrow{\text{EtO-}}$

b) ? $\xrightarrow{\text{EtO-}}$ $CH_3CH = CHCH_2CH_3$

c) ? $\xrightarrow{\text{EtO-}}$

Answer:
a) bromomethylcyclohexane
b) $CH_3CH_2CHBrCH_2CH_3$
c) $CH_3CH_2CH(CH_3)CH_2Br$
Section: 2

46. Write the product for the following reaction.

$\xrightarrow[\text{2) NaOEt, }\Delta]{\text{1) Br}_2\text{, h}\nu}$

Answer: 1-methylcyclohexene Section: 12

47. Devise methods by which the following molecule could be synthesized from cyclohexane.

Answer: Cl_2, hυ; NaOH, EtOH, heat Section: 12

48. Devise methods by which the following molecule could be synthesized. $(CH_3)_2C=CH_2$ from 2-methylpropane

Answer: Br_2, hυ; NaOH, EtOH, heat Section: 12

MULTIPLE CHOICE

49. When 2-bromo-3-methyl-1-phenylbutane is treated with sodium methoxide, why is the major product 3-methyl-1-phenyl-1-butene?

   a. S$_N$2 predominates over E2.
   b. E1 predominates over E2.
   c. The bulkiness of the methoxide results in the less substituted alkene.
   d. The newly formed double bond in this compound is conjugated with the phenyl ring.
   e. The less substituted alkene is always more stable than the more substituted alkene.

   Answer: d   Section: 2

50. Which of the following is <u>least likely</u> to be found in the product mixture which results when the alkyl iodide below is heated in water?

   a. I
   b. II
   c. III
   d. IV
   e. V

I.

II.

III.

IV.

V.

   Answer: d   Section: 8

SHORT ANSWER

51. Why is the E1 not a likely mechanism when 1-chloropentane is heated in ethanol?

Answer:
Primary halides cannot undergo E1 reactions because of the difficulty encountered in forming primary carbocations.

Section: 4

MULTIPLE CHOICE

52. Which of the following is <u>least likely</u> to be found in the product mixture which results when 2-iodopentane reacts with sodium ethoxide in ethanol?

a. 1-ethoxypentane
b. 2-ethoxypentane
c. (Z)-2-pentene
d. (E)-2-pentene
e. 1-pentene

Answer: a   Section: 5

53. What is the major product which results when (2R,3S)-2-chloro-3-phenylbutane is treated with sodium methoxide in methanol?

a. (E)-2-phenyl-2-butene
b. (Z)-2-phenyl-2-butene
c. (S)-3-phenyl-1-butene
d. (R)-3-phenyl-1-butene
e. (R)-2-methoxy-2-phenylbutane

Answer: b   Section: 5

SHORT ANSWER

54. Provide the structure of the major elimination product which results when the alkyl bromide below is heated in ethanol.

Answer:

Section: 6

55. Provide the structure of the major organic product of the following reaction.

Answer:

Section: 8

56. Can primary alkyl halides react by S_N2, S_N1, E2, and E1 mechanisms? Are any of these mechanisms prohibited? What conditions favor a particular mechanism?

Answer:
Only S_N2 and E2 pathways are available to primary halides. S_N1 and E1 cannot occur due to the instability of primary carbocations. S_N2 will typically predominate over E2 unless there is steric hindrance in the alkyl halide or nucleophile/base.

Section: 8

57. When nucleophile and leaving group are contained in the same molecule, is intermolecular or intramolecular reaction favored? Explain.

Answer:
The answer depends on the concentration of the bifunctional molecule and the size of the ring that will be formed in the intramolecular reaction. The intramolecular reaction has the advantage of reacting groups being tethered close together and therefore not having to wander through the solvent to find each other. As a result, a low concentration of reactant favors intramolecular reaction over intermolecular. Additionally, if the intramolecular reaction would form a five- or six-membered ring, it would be highly favored since smaller ring formations are retarded by angle strain and larger ring formations are retarded by less favorable entropy considerations.

Section: 11

MULTIPLE CHOICE

58. Which of the following represents the best preparation of 2-cyclopentenol from cyclopentane?

a. 1. Br_2, hυ 2. $NaOCH_3$ 3. NBS, CCl_4 4. H_2O
b. 1. NBS, CCl_4 2. Br_2, CCl_4 3. NaOH
c. 1. Br_2, hυ 2. H_2O_2
d. 1. NBS, CCl_4 2. $NaOCH_3$ 3. H_2O_2
e. 1. NBS, CCl_4 2. H_2O2 3. Br_2, CCl_4 4. $NaOCH_3$

Answer: a Section: 12

Chapter 11: Reactions at an sp3 Hybridized Carbon III

MULTIPLE CHOICE

1. Which of the following is the best method for preparing CH_3Br?

   a. $CH_3OH + Br^-$
   b. $CH_3OH + HBr$
   c. $CH_3OH + Br_2$
   d. $CH_3OH + NaBr$
   e. $CH_3OH + Br^+$

   Answer: b Section: 1

2. What is the major product of the following reaction?

$$CH_3 - \underset{\underset{CH_3}{|}}{\overset{\overset{CH_3}{|}}{C}} - \underset{\underset{OH}{|}}{CH}CH_3 + HBr \xrightarrow{\text{heat}}$$

   a.
$$CH_3 - \underset{\underset{CH_3}{|}}{\overset{\overset{CH_3}{|}}{C}} - \overset{\overset{Br}{|}}{CH}CH_3$$

   b.
$$CH_3 - \underset{\underset{CH_3}{|}}{\overset{\overset{CH_3}{|}}{C}} - CH_2\underset{\underset{Br}{|}}{CH_2}$$

   c.
$$CH_3 - \underset{\underset{Br}{|}}{\overset{\overset{CH_3}{|}}{C}} - \underset{\underset{CH_3}{|}}{CH}CH_3$$

   d.
$$CH_3 - \underset{\underset{CH_2Br}{|}}{\overset{\overset{CH_3}{|}}{C}} - CH_2CH_3$$

   e.
$$CH_3 - \underset{\underset{Br}{|}}{\overset{\overset{CH_3}{|}}{C}} - CH_2CH_2CH_3$$

   Answer: c Section: 1

3. What is the major product of the following reaction?

   a. I
   b. II
   c. III
   d. IV
   e. V

I.

II.

III.

IV.

V.

Answer: b Section: 1

4. Which of the following reagents is best used in the conversion of methyl alcohol to methyl chloride?

   a. Cl_2/CCl_4
   b. $Cl_2/h\upsilon$
   c. Cl^-
   d. $SOCl_2$
   e. $NaCl$

Answer: d Section: 2

5. What is the product of the following reaction?

$$CH_3-\overset{\overset{\displaystyle CH_2CH_3}{|}}{\underset{\underset{\displaystyle H}{|}}{}}-OH \quad + \text{ TsCl} \longrightarrow \quad ?$$

a.
$$CH_3-\overset{\overset{\displaystyle CH_2CH_3}{|}}{\underset{\underset{\displaystyle H}{|}}{}}-Cl$$

b.
$$CH_3-\overset{\overset{\displaystyle CH_2CH_3}{|}}{\underset{\underset{\displaystyle H}{|}}{}}-OTs$$

c.
$$Cl-\overset{\overset{\displaystyle CH_2CH_3}{|}}{\underset{\underset{\displaystyle H}{|}}{}}-CH_3$$

d.
$$TsO-\overset{\overset{\displaystyle CH_2CH_3}{|}}{\underset{\underset{\displaystyle H}{|}}{}}-CH_3$$

e.
$$CH_3-\overset{\overset{\displaystyle CH_2CH_3}{|}}{\underset{\underset{\displaystyle H}{|}}{}}-OCl$$

Answer: b Section: 3

6. What is the product of the following reaction?

$$CH_3-\overset{\overset{\displaystyle CH_2CH_3}{|}}{\underset{\underset{\displaystyle H}{|}}{}}-OH \xrightarrow{\text{TsCl}} \xrightarrow{Cl^-}$$

a.
$$CH_3-\overset{\overset{\displaystyle CH_2CH_3}{|}}{\underset{\underset{\displaystyle H}{|}}{}}-Cl$$

b.
$$CH_3-\overset{\overset{\displaystyle CH_2CH_3}{|}}{\underset{\underset{\displaystyle H}{|}}{}}-OTs$$

c.
$$Cl-\overset{\overset{\displaystyle CH_2CH_3}{|}}{\underset{\underset{\displaystyle H}{|}}{}}-CH_3$$

d.

$$CH_2CH_3$$
$$TsO \overline{} \Big|_H \overline{} CH_3$$

e.

$$CH_2CH_3$$
$$CH_3 \overline{} \Big|_H \overline{} OCl$$

Answer: c Section: 3

7. Which of the following alcohols dehydrates with the <u>fastest</u> rate?

a. I
b. II
c. III
d. IV
e. V

I. $CH_3\overline{}\bigcirc\overline{}OH$

II. CH_3, HO — cyclohexane

III. CH_3 — cyclohexane, HO

IV. $CH_3\overline{}$ cyclohexane, OH

V. cyclohexane $\overline{}CH_2OH$

Answer: b Section: 4

8. Which of the following alcohols gives a rearranged carbocation when dehydrated?

   a. I
   b. II
   c. III
   d. IV
   e. V

I. CH_3CHCH_3
       |
      OH

II. $CH_3CHCH_2CH_3$
      |
     OH

III. (cyclohexane ring)—OH

IV. CH_3 (on cyclohexane ring)—OH

V. CH_3
       |
  CH_3-C-CH_3
       |
     OH

Answer: d   Section: 4

9. What is the product of the following reaction?

   a. I
   b. II
   c. III
   d. IV
   e. V

$$\text{[cyclobutane]}-\underset{\underset{OH}{|}}{CHCH_3} \quad \xrightarrow[\text{heat}]{H_2SO_4} \quad ?$$

I.

II.

III. [cyclobutane]—CH=CH₂

IV. [cyclobutane]=CHCH₃

V. [cyclobutane]—CH₂CH₃

Answer: a   Section: 4

233

10. What is the <u>major</u> product of the following reaction?

    a. I
    b. II
    c. III
    d. IV
    e. V

Answer: b Section: 4

11. What is the major product of the following reaction?

    a. I
    b. II
    c. III
    d. IV
    e. V

$$CH_3CH_2OH \ + \ \bigcirc\!\!-\!SO_2Cl \longrightarrow \ ?$$

    I. $CH_2{=}CH_2$

    II. CH_3CH_2Cl

    III. $CH_3CH_2SO_2Cl$

    IV. $CH_3CH_2SO_2{-}\bigcirc$

    V. $CH_3CH_2OSO_2{-}\bigcirc$

Answer: e Section: 3

12. What are the major products from the following reaction?

a. I
b. II
c. III
d. IV
e. V

I.

II.

III.

IV.

V.

Answer: b Section: 5

13. What is/are the product(s) from the following reaction?

    a. I
    b. II
    c. III
    d. IV
    e. V

  I. $HOCH_2CH_2I$

  II. CH_3OH CH_3I

  III.

  IV.

  V. CH_3OCH_2I

Answer: a Section: 6

14. What is the <u>major</u> product from the following reaction?

   a. I
   b. II
   c. III
   d. IV
   e. V

I.

II. $HOCH_2CHCH_3$
                $|$
                Br

III. $BrCH_2CHCH_3$
                  $|$
                  OH

IV.

V. $HOCH_2CH_2CH_2Br$

Answer: b   Section: 6

15. What is the major product for the following reaction?

    a. I
    b. II
    c. III
    d. IV
    e. V

$$\underset{\underset{CH_3}{|}}{\overset{O}{\triangle}}CH_3 \quad \xrightarrow[CH_3OH]{H^+} \quad ?$$

I.   CH₃OCH₂CH₂CH₂CH₂OH

II. $\underset{\underset{CH_3}{|}}{HOCHCHOCH_3}\overset{\overset{CH_3}{|}}{}$

III.

IV. $\overset{\overset{CH_3}{|}}{\underset{\underset{OCH_3}{|}}{HOCH_2CCH_3}}$

V. $\overset{\overset{CH_3}{|}}{\underset{\underset{OH}{|}}{CH_3OCH_2CCH_3}}$

Answer: d   Section: 6

16. What is the <u>major</u> product for the following reaction?

    a. I
    b. II
    c. III
    d. IV
    e. V

I. $CH_3OCH_2CH_2CH_2CH_2OH$

II. $\underset{\underset{CH_3}{|}}{HOCHCHOCH_3}$ with CH_3 above

III. (epoxide structure with H, $O^{\oplus}$, CH_3, CH_3) $\overset{\ominus}{O}CH_3$

IV. $HOCH_2\overset{\overset{CH_3}{|}}{\underset{\underset{OCH_3}{|}}{C}}CH_3$

V. $CH_3OCH_2\overset{\overset{CH_3}{|}}{\underset{\underset{OH}{|}}{C}}CH_3$

Answer: e Section: 6

17. Which of the following is the strongest base?

    a. HOMgBr
    b. H_2O
    c. CH_4
    d. CH_3OH
    e. CH_3MgBr

Answer: e Section: 8

18. What is the product of the following reaction?

    a. I
    b. II
    c. III
    d. IV
    e. V

I.

II.

CH₃CHCH₂—⬡
    |
   OH

III.   HOCH₂CH₂—⬡

IV.   HOCH₂CH₂CH₂—⬡

V.

Answer: d Section: 8

19. What is the product of the following reaction sequence?

a. I
b. II
c. III
d. IV
e. V

I.   Ph$-$C≡CCH$_3$

II.   Ph$-$C≡CCH$_2$CH$_2$OCH$_3$

III.   Ph$-$C≡CCH$_2$OCH$_2$CH$_3$

IV.   Ph$-$C≡CCH$_2$CH$_2$OH

V.   Ph$-$C≡CCH$_2$CH$_2$CH$_2$OH

Answer: d   Section: 8

20. What is the product of the following reaction?

    a. I
    b. II
    c. III
    d. IV
    e. V

Answer: c Section: 8

21. Which of the following is phenylthiol?

    a. I
    b. II
    c. III
    d. IV
    e. V

I.

II.

III.

IV.

V.

Answer: b   Section: 10

22. Which of the following is the strongest acid?

    a. CH_3NH
    b. CH_3OH
    c. CH_3SH
    d. CH_3OCH_3
    e. CH_3Cl

Answer: c   Section: 10

23. Which of the following is a quaternary ammonium ion?

a.
$$R-\overset{\displaystyle R}{\underset{\displaystyle }{C}}-R$$

b.
$$R-\overset{\displaystyle \overset{\oplus}{R}}{\underset{\displaystyle R}{C}}-R$$

c.
$$R-\overset{\displaystyle R}{\underset{\displaystyle \cdot\cdot}{N}}-R$$

d.
$$R-\overset{\displaystyle R}{\underset{\displaystyle R}{N}}-R$$

e.
$$R-\overset{\displaystyle \overset{\oplus}{R}}{\underset{\displaystyle R}{N}}-R$$

Answer: e Section: 11

24. What is the major alkene formed in the following Hofmann elimination?

$$CH_3CH_2CH_2\overset{\overset{\displaystyle \overset{\ominus}{OH}}{\displaystyle \overset{CH_3}{\mid}\oplus}}{\underset{\displaystyle CH_3}{N}}\!\!-\!\!\!-\!\!\!-\overset{\displaystyle \overset{CH_3}{\mid}}{\underset{\displaystyle CH_3}{C}}CH_3 \xrightarrow{\text{heat}} ?$$

a. $CH_3CH=CH_2$

b. $CH_3-\overset{\displaystyle }{\underset{\displaystyle \overset{\|}{CH_2}}{C}}-CH_3$

c. $CH_3CH_2CH=CH_2$

d. $CH_3CHCH=CH_2$
$\qquad \ \ |$
$\qquad \ \ CH_3$

e. $CH_3CH_2C=CH_2$
$\qquad \quad \ \ |$
$\qquad \quad \ \ CH_3$

Answer: a Section: 11

25. What is the major alkene formed in the following Hofmann elimination?

a. I
b. II
c. III
d. IV
e. V

I.

II.

III.

IV.

V.

Answer: c Section: 11

26. What is the product of the following reaction?

a. I
b. II
c. III
d. IV
e. V

I.

II.

III.

IV.

V.

Answer: e Section: 11

27. What is the <u>major</u> alkene formed in the following reaction?

    a. I
    b. II
    c. III
    d. IV
    e. V

Answer: c Section: 11

28. What is the <u>major</u> alkene produced in the following Cope elimination?

    a. $CH_2=CH_2$
    b. $CH_3CH=CH_2$
    c. $CH_3CH_2N=CH_2$
    d. $CH_3CH_2N=O$
    e. $CH_3CH_2CH_2N=O$

Answer: a Section: 11

SHORT ANSWER

29. Which of the following alcohols, when heated with H_2SO_4, will undergo dehydration more rapidly? Explain.

a) [cyclohexyl-CH–CH$_3$ / OH structure]   or   b) [phenyl-CH–CH$_3$ / OH structure]

Answer:
b) will undergo dehydration more rapidly because it produces a secondary benzylic carbocation, which is more stable than secondary carbocation.

Section: 4

30. A compound (A) $C_4H_{10}O$ is optically active. (A) reacts with HCl and $ZnCl_2$ (Lucas test) to form a cloudy product within 10 minutes. Deduce the structure of (A).

Answer:

One of the enantiomers of $CH_3CH_2\overset{\overset{\displaystyle H}{|}}{\underset{\underset{\displaystyle CH_3}{|}}{C}}$—OH

Section: 1

31. The following alkyl halide could not be used to form a Grignard reagent. Explain.

$$HO-CH_2CH_2-Br$$

Answer:
Since the alkyl halide contains an acidic proton on the oxygen, the Grignard reagent could not be formed.

Section: 8

249

32. Show how the following synthesis could be carried out.

Answer:

Section: 8

33. Starting with isopropyl alcohol as the only organic material, synthesize isopropyl propyl ether.

$$CH_3CH_2CH_2OCHCH_3$$
(with CH_3 branch)

Answer:

$$CH_3CHCH_3 \xrightarrow[\text{heat}]{H_2SO_4} CH_3CH= CH_2 \xrightarrow[\text{Peroxide}]{HBr} CH_3CH_2CH_2Br$$
(with OH on the first structure)

$$+ \longrightarrow CH_3CH_2CH_2OCH \begin{smallmatrix} CH_3 \\ CH_3 \end{smallmatrix}$$

$$CH_3CHCH_3 \xrightarrow{Na} CH_3CHCH_3$$
(OH) (O⁻)

Section: 4

34. Propose a mechanism for the following reaction:

$$CH_2CH_2CH_2CH_2(OH)(OH) \xrightarrow{\ H^+\ } \text{(tetrahydrofuran)} + H_2O$$

Answer:

Section: 1

35. Propose a mechanism for the following reaction:

$$\text{(tetrahydrofuran)} + HBr \xrightarrow{\ heat\ } CH_2CH_2CH_2CH_2(Br)(Br)$$

Answer:

Section: 5

36. Propose a detailed, step-by-step mechanism for the reaction pathway shown below.

   Answer:

Section: 4

37. Propose a detailed, step-by-step mechanism for the reaction pathway shown below.

   Answer:

Section: 4

38. Draw all likely products of the following reaction and circle the product you expect to predominate.

2-pentanol $\xrightarrow{H_2SO_4}$

Answer:

Section: 4

39. Draw all likely products of the following reaction and circle the product you expect to predominate.

Answer:

Section: 4

40. Draw all likely products of the following reaction and circle the product you expect to predominate.

Answer:

Section: 4

41. Provide the structure of the major organic product in the reaction below.

1. Ag$_2$O
2. Δ
3. CH$_3$I
4. Ag$_2$O
5. Δ

Answer:

Section: 11

42. Provide the structure of the major organic product in the reaction below.

1. CH$_3$I (excess)
2. Ag$_2$O
3. Δ

Answer:

Section: 11

43. Provide the structure of the major organic product in the reaction below.

1. CH$_3$I (excess)
2. Ag$_2$O
3. Δ

Answer:

Section: 11

44. Provide the structure of the major organic product in the reaction below.

$$\xrightarrow[\text{2. } \Delta]{\text{1. } H_2O_2}$$

Answer:

Section: 11

45. Provide the structure of the major organic product in the reaction below.

$$\xrightarrow[\text{2. } \Delta]{\text{1. } H_2O_2}$$

Answer:

Section: 11

46. How would one use a Gringnard-based synthesis to accomplish the following transformation?

benzyl bromide ($PhCH_2Br$) to 3-phenyl-1-propanol

Answer:
1. Mg, Et_2O
2. ethylene oxide (oxirane)
3. H_3O^+

Section: 8

47. Provide the reagents necessary to accomplish the following transformation.

Answer:
1. Mg, Et_2O
2. D_2O

Section: 8

48. Draw the tosylate ion and explain why it is a particularly good leaving group.

Answer:
Extensive resonance stabilization of this ion makes it a particularly good leaving group.

Section: 3

49. Why are alcohols typically poor electrophiles?

Answer:
The hydroxide ion, a relatively strong base, is a very poor leaving group.

Section: 1

50. Provide the structure of the major organic product in the reaction below.

$PhCH_2CH_2OH$ $\xrightarrow[\text{2. NaCN, acetone}]{\text{1. TsCl, pyridine}}$

Answer: $PhCH_2CH_2CN$ Section: 3

51. Provide the structure of the major organic product in the reaction below.

1. TsCl, pyridine
2. NaOEt, Δ

Answer:

Section: 4

MULTIPLE CHOICE

52. When (R)-2-butanol is treated with TsCl in pyridine, the product formed is:

   a. an achiral compound.
   b. a mixture of diastereomers.
   c. a racemic mixture.
   d. a single enantiomer.
   e. none of the above.

Answer: d   Section: 3

SHORT ANSWER

53. Predict the major product of the reaction below and provide a stepwise mechanism which accounts for its formation.

$CH_3(CH_2)_6CH_2OH$ + HBr $\longrightarrow$

Answer:

Section: 1

MULTIPLE CHOICE

54. Which of the following alcohols will react most rapidly with the Lucas reagent (HCl, $ZnCl_2$)?

    a. $(CH_3)_3COH$
    b. $CH_3CH_2CH_2CH_2OH$
    c. $CH_3CHOHCH_2CH_3$
    d. $(CH_3)_2CHCH_2OH$

    Answer: a   Section: 1

SHORT ANSWER

55. Provide the structure of the major organic product in the reaction below.

    Answer:

    Section: 1

56. Provide the structure of the major organic product in the reaction below.

    Answer:

    Section: 1

57. Provide the structure of the major organic product in the reaction below.

$(CH_3)_2CHCH_2OH \xrightarrow{\text{PBr}_3}$

Answer: $(CH_3)_2CHCH_2Br$ Section: 2

58. Provide the major organic product in the reaction below.

CH_3CH_2, CH_3 O epoxide $+$ NaOCH$_3$ $\longrightarrow$

Answer:

CH_3 CH_3 CH_3 OH OCH_3

Section: 6

59. Provide the major organic product in the reaction below.

$\xrightarrow{\text{HBr}}$

Answer:

Section: 6

60. Provide the major organic product in the reaction below.

$\xrightarrow[\text{2. H}_3\text{O}^+]{\text{1. CH}_3\text{CH}_2\text{MgBr}}$

Answer:

OH CH$_3$

Section: 8

259

MULTIPLE CHOICE

61. When 1-pentanol is heated with HCl/ZnCl$_2$, 1-chloropentane is the major organic product. This reaction proceeds through an _____ mechanism, and _____ is produced as a byproduct.

    a. S$_N$1, H$_2$O
    b. S$_N$2, H$_2$O
    c. S$_N$1, H$_2$
    d. S$_N$2, H$_2$
    e. E2, H$_2$

    Answer: b   Section: 1

SHORT ANSWER

62. Name and provide the structure of the major organic product which results when (R)-3,3-dimethyl-2-butanol is heated in the presence of HBr.

    Answer: 2-bromo-2,3-dimethylbutane

    Section: 1

63. Provide the structure of the major organic product of the following reaction. Explain the stereochemistry which results in this product.

    H$_3$CH$_2$CH$_2$C — CH$_3$, H, OH $\xrightarrow[\text{2. NaCN}]{\text{1. PBr}_3\text{, pyridine}}$

    Answer:
    Each of the two steps proceeds via an SN2 mechanism. Two stereochemical inversions at the chirality center occur; thus, the product appears to have retained the original stereochemistry.

    H$_3$CH$_2$CH$_2$C — CH$_3$, H, OH

    Section: 2

MULTIPLE CHOICE

64. Which of the following is <u>least likely</u> to be found in the product mixture which results when 2,2-dimethyl-3-pentanol is heated in phosphoric acid?

    a. (E)-4,4-dimethyl-2-pentene
    b. (Z)-4,4-dimethyl-2-pentene
    c. 2,3-dimethyl-2-pentene
    d. 2,3-dimethyl-1-pentene
    e. 4,4-dimethyl-1-pentene

    Answer: e   Section: 4

65. When dipentyl ether is treated with excess HI, through what type of mechanism(s) does the major product result?

    a. S_N2
    b. S_N1
    c. both S_N1 and S_N2
    d. E1
    e. E2

    Answer: a   Section: 5

SHORT ANSWER

66. Provide the structure of the major organic product which results when naphthalene oxide undergoes rearrangement in the presence of water. Explain the regioselectivity of this rearrangement.

    Answer:
    Naphthalene oxide can rearrange to form either 1-naphthol or 2-naphthol. The carbocation leading to 1-naphthol is more stable because more electron delocalization is present in which aromaticity is retained in the benzene ring.

    Section: 7

67. How are certain arene oxides believed to cause cancer?

Answer:
2'-Deoxyguanosine, a component of DNA, is a nucleophile and reacts with certain arene oxides to form a covalent bond. Once this part of DNA has formed a covalent bond to another compound, the structure and replication of the DNA is disrupted. The genetic code can no longer be transcribed properly, and this can lead to cancerous mutations.
Section: 7

MULTIPLE CHOICE

68. Which of the following sequences of reactions would be best when converting cyclohexene oxide into propylcyclohexane?

a. 1. CH_3CCNa 2. H_3O+
b. 1. $CH_3CH_2CH_2MgBr$ 2. H_2SO_4, Δ 3. H_2, Pt
c. 1. H_3O^+ 2. CH_3CCNa
d. 1. $CH_3CH_2CH_2Li$ 2. HBr
e. 1. H_3PO_4, Δ 2. $CH_3CH_2CH_2MgBr$ 3. H^+

Answer: b   Section: 8

69. What is the major organic product of the following reaction sequence?

$$CH_3C \equiv CCH_3 \quad \xrightarrow[\text{2. } (CH_3CH_2CH_2)_2CuLi]{\text{1. HBr}}$$

a. (E)-3-methyl-2-hexene
b. 2-heptyne
c. 5-decyne
d. (Z)-2-bromo-2-hexene
e. (E)-2-bromo-2-hexene

Answer: a   Section: 8

SHORT ANSWER

70. Provide the sequence of reagents needed to carry out the following conversion.

$$CH_2C\equiv CH \quad \xrightarrow{\hspace{4cm}} \quad CH_3CH_2CH_2CH_2CH_2OH$$

Answer:
1. $NaNH_2$
2. ethylene oxide
3. H_3O^+
4. H_2 (excess), Pt
Section: 8

Chapter 12: Mass Spectrometry, Infrared Spectroscopy, and UV/Vis

MULTIPLE CHOICE

1. Which of the following statements best explains the information we can gain from mass spectrometry?

   a. It allows us to determine the number of protons in a compound.
   b. It allows us to determine the kinds of functional groups in a compound.
   c. It allows us to determine the molecular weight and the mass of some fragments of a compound.
   d. It allows us to determine the presence and nature of a carbocation in the compound.
   e. It allows us to determine the presence and nature of a free radical in the compound.

   Answer: c Section: 98

2. Which of the following statements best explains the information we can gain from infrared spectroscopy?

   a. It allows us to determine the number of protons in a compound.
   b. It allows us to determine the kinds of functional groups in a compound.
   c. It allows us to determine the molecular weight and the mass of some fragments of a compound.
   d. It allows us to determine the presence and nature of a carbocation in the compound.
   e. It allows us to determine the presence and nature of a free radical in the compound.

   Answer: b Section: 98

3. Which of the following molecular changes is necessary for mass spectrometry to occur?

   a. Excitation of an electron from the ground state to higher energy state
   b. Change of alignment of an electron in a magnetic field
   c. Change of alignment of a proton in a magnetic field
   d. Loss of an electron
   e. Molecular vibration

   Answer: d Section: 1

4. Which of the following statements is true about what a molecular ion is?

    a. A compound that lost a pair of electrons
    b. A compound that gained a pair of electrons
    c. A compound that gained one electron
    d. A compound that lost one electron
    e. A compound that carries a free radical and a negative charge

Answer: d   Section: 1

5. Which of the following statements best describes the meaning of the following species:

$$[CH_3CH_2CH_3]\cdot{}^+ \ ?$$

    a. It is the molecular ion of propane.
    b. It is the parent ion of propane.
    c. It is the radical cation of propane.
    d. The m/z value is 44.
    e. All of the above.

Answer: e   Section: 1

6. Which of the following is true about the base peak in mass spectrometry?

    a. The m/z value equals the molecular weight of the compound.
    b. The m/z value corresponds to a very stable carbocation.
    c. It has the largest peak height in the spectrum.
    d. It has the highest m/z value of all the peaks in the spectrum.
    e. The base peak is assigned a relative abundance equal to that of the parent ion.

Answer: c   Section: 2

7. Which of the following is **not** true about the M+1 peak?

    a. It is one m/z unit higher than the base peak.
    b. It is one m/z unit higher than the molecular ion peak.
    c. It is one m/z unit higher than the parent ion peak.
    d. It occurs because there are two naturally occurring isotopes of carbon.
    e. This means that the number of carbon atoms in a compound can be calculated if the relative abundance of both the M and M+1 peaks is known.

Answer: a   Section: 3

8. Which of the following statements best explains how a hydrocarbon can show an M+2 peak in mass spectrometry?

   a. From ^{13}C and ^{1}H
   b. From ^{12}C and ^{2}H
   c. From a single ^{13}C
   d. From a single ^{3}H

   e. From two ^{13}C's

   Answer: e Section: 3

9. Which of the following m/z values is the base peak for benzyl alcohol?

   a. 77
   b. 108
   c. 91
   d. 17
   e. 52

   Answer: c Section: 5

10. Which of the following m/z values is the molecular ion for 2-butanone?

   a. 15
   b. 29
   c. 43
   d. 57
   e. 72

   Answer: e Section: 1

11. Which of the following compounds exhibits the pattern of m/z values shown below?

   41, 43, 57, 87, 101, 116

   a. propylbromide
   b. isopropyl bromide
   c. sec-butyl isopropyl ether
   d. 2-hexanol
   e. 2-butanone

   Answer: c Section: 5

12. Which of the following statements is <u>not</u> true about electromagnetic radiation?

    a. The velocity of light is directly proportional to the energy.
    b. All molecules absorb electromagnetic radiation at some frequency.
    c. Frequency is inversely proportional to wavelength.
    d. Energy is directly proportional to frequency.
    e. Energy is inversely proportional to wavelength.

    Answer: a Section: 6

13. Which of the following regions of the electromagnetic spectrum has the greatest energy per photon?

    a. Visible
    b. Microwave
    c. Radio
    d. Ultraviolet
    e. Infrared

    Answer: d Section: 6

14. Which of the following normally occurs in a molecule when a photon of infrared light is absorbed?

    a. An electron moves to an orbital of higher potential energy.
    b. The vibration energy increases.
    c. An electron changes alignment in a magnetic field.
    d. The molecule gains an electron.
    e. The molecule loses an electron.

    Answer: b Section: 6

15. How many centimeters (cm) are there in one micrometer (μm)?

    a. 10^{-6}
    b. 10^{-3}
    c. 10^{-5}
    d. 10^{-8}
    e. 10^{-4}

    Answer: e Section: 6

16. Which of the following are considered to be bending (in-plane) vibrations?

   a. scissoring and wagging
   b. scissoring and twisting
   c. rocking and wagging
   d. rocking and twisting
   e. scissoring and rocking

   Answer: e Section: 7

17. Which of the following solvents is best used in infrared spectroscopy?

   a. I
   b. II
   c. III
   d. IV
   e. V

   I.   Water, H_2O

   II. Carbon tetrachloride, CCl_4

   III. Methanol, CH_3OH

   IV. Ethanol, CH_3CH_2OH

   V.   Benzene,

   Answer: b Section: 7

18. Which of the infrared regions is considered to be the <u>fingerprint</u> region?

   a. $4000 cm^{-1}$ - $1000 cm^{-1}$
   b. $4000 \mu m$ - $1000 \mu m$
   c. $2200 \mu m$ - $1000 \mu m$
   d. $1000 cm^{-1}$ - $400 cm^{-1}$
   e. $1000 \mu m$ - $400 \mu m$

   Answer: d Section: 7

19. Which of the following wavenumbers corresponds to the bond shown below?
   $$C \equiv C$$

   a. $1650 \ cm^{-1}$
   b. $2100 \ cm^{-1}$
   c. $1100 \ cm^{-1}$
   d. $3300 \ cm^{-1}$
   e. $2850 \ cm^{-1}$

   Answer: b Section: 10

20. Which of the following carbonyl groups exhibits the highest wavenumber in infrared spectroscopy?

   a. I
   b. II
   c. III
   d. IV
   e. V

Answer: c Section: 10

21. Which of the following carbon-hydrogen bonds exhibits the lowest wavenumber for a C—H stretch in infrared spectroscopy?

   a. I
   b. II
   c. III
   d. IV
   e. V

   I.

   II. $\overset{\overset{\textstyle O}{\|}}{C}$—H

   III. C—C—H

   IV. C=C—H

   V. C≡C—H

   Answer: b Section: 11

22. Which of the following compounds has a vibration that is infrared inactive?

   a. CH_3—C≡C—H
   b. H_2O
   c.
      $\overset{\overset{\textstyle O}{\|}}{CH_3-C-CH_3}$
   d. CO_2
   e. CO

   Answer: d Section: 14

23. Which of the following functional groups will exhibit <u>no</u> IR absorption at 1630 - 1780 cm^{-1} <u>or</u> at 3200 - 3550 cm^{-1}?

   a. An alcohol
   b. An amide
   c. A ketone
   d. An aldehyde
   e. An ether

   Answer: e Section: 13

24. In which of the following molecules is the presence of the IR peak for $C=C$ not observed?

a.
$$\underset{H}{\overset{H}{\diagdown}}\ \underset{H}{\overset{H}{\diagup}}$$
$$C=C$$

b.
$$\underset{Cl}{\overset{H}{\diagdown}}\ \underset{Cl}{\overset{Cl}{\diagup}}$$
$$C=C$$

c.
$$\underset{H}{\overset{CH_3}{\diagdown}}\ \underset{CH_2CH_3}{\overset{H}{\diagup}}$$
$$C=C$$

d.
$$\underset{CH_3}{\overset{CH_3}{\diagdown}}\ \underset{H}{\overset{H}{\diagup}}$$
$$C=C$$

e.
$$\underset{CH_3}{\overset{H}{\diagdown}}\ \underset{CH_3}{\overset{H}{\diagup}}$$
$$C=C$$

Answer: a   Section: 14

25. Predict the molecular formula of the compound represented below based on the MS data given.

m/z	Intensity
84	10.00
85	0.56
86	0.04

a. C_6H_{12}
b. C_5H_{24}
c. $C_4H_6O_2$
d. $C_3H_8O_2$
e. C_5H_8O

Answer: e   Section: 3

26. Which of the following characterizes the <u>unusually</u> intense peak of alkyl chlorides in MS spectrometry?

    a. M + 1 peak
    b. M + 2 peak
    c. Base peak
    d. Parent peak
    e. None of the above

    Answer: b Section: 3

27. Which of the following is the base peak for the compound below?

    a. 77
    b. 92
    c. 15
    d. 57
    e. 43

    Answer: d Section: 5

28. Which of the following structures will give a base peak of 43 in mass spectrometry?

    a. $CH_3CH_2CHCH_3$
          |
          CH_2CH_3
    b. $CH_3CH_2CH_2CH_2CH_2CH_3$
    c.
    d. CH_3
            |
         $CH_3CCH_2CH_3$
            |
            CH_3
    e. None of the above

    Answer: c Section: 5

SHORT ANSWER

29. What does m/z stand for and what does it mean?

Answer:
The m stands for the mass and z for the charge. Since most fragments have a charge of +1, m/z is the molecular weight of the fragment.
Section: 1

30. How could you distinguish the mass spectrum of 2,2-dimethylpropane from that of isopentane?

Answer:
The loss of a methyl from 2,2-dimethylpropane forms a stable tertiary carbocation, while loss of a methyl from isopentane forms a less stable secondary carbocation. So the M-15 peak will be greater for 2,2-dimethylpropane.
Section: 2

31. Is it possible to have an M + 2 peak in mass spectrometry? Explain.

Answer:
A mass spectrum can show an M + 2 peak for a compound with a relatively abundant isotope two mass units greater than the most common isotope (such as oxygen, chlorine, and bromine) or for a compound that contains two isotopes that are one mass unit heavier than the most abundant. The presence of a relatively large M + 2 peak is evidence for a compound with chlorine or bromine. For instance, if M + 2 is 1/3 the height of the molecular ion, the compound contains chlorine. If M + 2 is about the same height as that of the molecular ion, the compound contains bromine.
Section: 3

32. The mass spectrum of an unknown compound has a molecular ion peak with a relative abundance of 43.27% and an M + 1 peak with a relative abundance of 3.81%. How many carbon atoms are in the compound?

Answer:
\# of carbon atoms = $\dfrac{\text{relative abundance of M + 1}}{.011 \cong \text{(relative abundance of M)}}$

$\dfrac{3.81}{.011 \cong 43.27}$ = 8 carbons

Section: 3

33. Show the m/z values of the molecular ion and 5 likely fragments for the compound ethyl ether,

$$CH_3CH_2OCH_2CH_3$$

Answer: 74, 59, 45, 44, 30, 29, 15 Section: 5

34. Does the position of the stretch of the oxygen-hydrogen bond in alcohols depend on the concentration of the alcohol?

Answer:
Yes, the more concentrated the alcohol, the more likely it is for the alcohol to form intermolecular hydrogen bonding. So the OH group of a concentrated (hydrogen-bonded) solution absorbs at a lower wavenumber than that of the diluted solution.

Section: 10

MULTIPLE CHOICE

35. Absorption of UV-visible energy by a molecule results in:

   a. vibrational transitions
   b. electronic transitions
   c. rotational transitions
   d. nuclear transitions
   e. none of the above

Answer: b Section: 16

36. Which compound would be expected to show intense IR absorption at 3300 cm^{-1}?

   a. CH_3CCCH_3
   b. butane
   c. 1-butene
   d. CH_3CH_2CCH

Answer: d Section: 11

37. Which compound would be expected to show intense IR absorption at 2710 and 1705 cm^{-1}?

   a. $CH_3COCH_2CH_3$
   b. $PhCOCH_3$
   c. PhCHO
   d. $CH_2{=}CHCOCH_3$

Answer: c Section: 10

38. Which compound would be expected to show intense IR absorption at 2250 cm^{-1}?

    a. $CH_3CH_2CH_2CO_2H$
    b. $CH_3CH_2CH_2CN$
    c. $CH_3CH_2CH_2CH_2OH$
    d. $CH_3CH_2CH_2CONH_2$

    Answer: b Section: 10

SHORT ANSWER

39. Deduce a possible structure for the compound with the IR absorptions below.

    C_3H_3Br: 3300, 2900, 2100 cm^{-1}

    Answer: $HCCCH_2Br$ Section: 15

40. Deduce a possible structure for the compound with the IR absorptions below.

    C_3H_5N: 3000, 2250 cm^{-1}

    Answer: CH_3CH_2CN Section: 15

41. Deduce a possible structure for the compound with the IR absorptions below.

    C_5H_8O: 3000, 1750 cm^{-1}

    Answer: cyclopentanone Section: 15

42. Deduce a possible structure for the compound with the IR absorptions below.

    C_4H_8O: 3000, 2715, 1715 cm^{-1}

    Answer: $CH_3CH_2CH_2CHO$ Section: 15

43. How could IR spectroscopy be used to distinguish between the following pair of compounds?

    $CH_3OCH_2CH_3$ and $CH_3CH_2CH_2OH$

    Answer:

    O-H stretch at 3300 cm^{-1}

    Section: 15

44. How could IR spectroscopy be used to distinguish between the following pair of compounds?

$HOCH_2CH_2CHO$ and $CH_3CH_2CO_2H$

Answer:
Carboxylic acid will have a very broad O-H absorption; aldehyde has a characteristic C-H stretch
Section: 15

MULTIPLE CHOICE

45. Ethyne (HCCH) does not show IR absorption in the region 2000-2500 cm^{-1} because:

a. C-H stretches occur at lower energies.
b. CC stretches occur at about 1640 cm^{-1}.
c. there is no change in the dipole moment when the CC bond in ethyne stretches.
d. there is a change in the dipole moment when the CC bond in ethyne stretches.

Answer: c Section: 14

SHORT ANSWER

46. Describe the molecular ion region in the mass spectrum of CH_3CH_2Br.

Answer:

The natural abundance of the isotopes ^{79}Br and ^{81}Br are about the same; therefore, there will be peaks of equal intensity at m/z 108 and 110.
Section: 3

47. 2-Methylhexane shows an intense peak in the mass spectrum at m/z = 43. Propose a likely structure for this fragment.

Answer: $(CH_3)_2CH+$ Section: 2

MULTIPLE CHOICE

48. Which compound would be expected to show intense IR absorption at 1715 cm^{-1}?

    a. $CH_3CH_2CO_2H$
    b. 1-hexene
    c. 2-methylhexane
    d. $CH_3CH_2CH_2NH_2$

    Answer: a   Section: 8

49. Which compound would be expected to show intense IR absorption at 1746 cm^{-1}?

    a. $CH_3CH_2OCH_2CH_3$
    b. $CH_3CO_2CH_3$
    c. CH_3CH_2CCH
    d. $CH_3CH_2SCH_3$

    Answer: b   Section: 8

SHORT ANSWER

50. An infrared wavelength of 4.48μm is equivalent to a wavenumber of _____ cm^{-1}.

    Answer: 2230   Section: 6

51. Which has the higher speed in a vacuum, ultraviolet or infrared light?

    Answer: They have the same speed   Section: 6

52. Which region of the electromagnetic spectrum, IR or UV, contains photons of the higher energy?

    Answer: UV   Section: 6

53. Which region of the electromagnetic spectrum, radio or visible, is characterized by waves of shorter wavelength?

    Answer: visible   Section: 6

54. Which region of the electromagnetic spectrum, IR or X-ray, is characterized by waves of lower frequency?

    Answer: IR   Section: 6

55. Which has a lower characteristic stretching frequency, the C-H or C-D bond? Explain briefly.

    Answer: C-D; heavier atoms vibrate more slowly   Section: 10

56. Which has a lower characteristic stretching frequency, the C=O bond or the C-O bond? Explain briefly.

    Answer:
    Stronger bonds are generally stiffer, thus requiring more force to stretch or compress them. The C-O bond is the weaker of the two and hence has the lower stretching frequency.
    Section: 10

57. Describe the fate of a molecule from introduction to detection in a mass spectrometer.

    Answer:
    Upon introduction, sample molecules are ionized by an electron beam passing through a vacuum chamber. The resulting radical cation and fragment cations are accelerated into the flight tube of the magnet by a negatively charged plate. In the flight tube, the path of the ions is bent by the existing magnetic field. For a given magnetic field strength, only one m/z will be bent in such a way that its path matches the curvature of the tube and reaches the detector. The entire m/z range is scanned by varying the strength of the magnetic field.
    Section: 1

58. What technique can be used to determine the molecular formula of a compound?

    Answer: High resolution mass spectrometry   Section: 4

MULTIPLE CHOICE

59. The mass spectra of alcohols often fail to exhibit detectable M peaks but instead show relatively large _____ peaks.

    a. M+1
    b. M+2
    c. M-16
    d. M-17
    e. M-18

    Answer: e   Section: 5

SHORT ANSWER

60. How could IR spectroscopy be used to distinguish between the following pair of compounds?

$(CH_3)_3N$ and $CH_3NHCH_2CH_3$

Answer:

N-H absorption near 3300 cm^{-1}

Section: 13

MULTIPLE CHOICE

61. In the UV-visible spectrum of (E)-1,3,5-hexatriene, the lowest energy absorption corresponds to:

a. a π to π* transition.
b. a σ to π* transition.
c. a π to σ* transition.
d. a σ to σ* transition.
e. a σ to π transition.

Answer: a   Section: 16

62. Which of the following compounds absorbs the longest wavelength of UV-visible light?

a. (E)-2-butene
b. (Z)-2-butene
c. 1-hexene
d. (Z)-1,3-hexadiene
e. (E)-1,3,5-hexatriene

Answer: e   Section: 16

63. Which of the following methods is most suitable for studying conjugation in molecules?

a. Infrared
b. NMR
c. Mass spectrometry
d. X-ray
e. UV Visible

Answer: e   Section: 16

64. Which of the following compounds exhibits the highest λ_{max} in UV spectra?

   a. I
   b. II
   c. III
   d. IV
   e. V

   I.

   II.

   III.

   IV.

   V.

   Answer: d Section: 16

65. Which color is observed when a sample absorbs light at 650 nm?

   a. yellow
   b. orange
   c. red
   d. green
   e. blue

   Answer: e Section: 19

66. What m/z value would you predict for the most stable fragment peak in the mass spectrum of 4-ethylheptane?

   a. 128
   b. 127
   c. 113
   d. 99
   e. 85

   Answer: b Section: 2

67. Which of the following m/z values is least intense in the mass spectrum of 1-chloropropane?

    a. 80
    b. 79
    c. 78
    d. 65
    e. 63

    Answer: b   Section: 5

SHORT ANSWER

68. Provide the structure of the two ions which result when the molecular ion of 2-methoxypentane undergoes fragmentation by α-cleavage.

    Answer:

    $CH_3CH_2CH_2CH \overset{+}{=} \overset{..}{O}CH_3$ $CH_3CH \overset{+}{=} \overset{..}{O}CH_3$

    Section: 5

69. Provide the structure of the species which results when the molecular ion of 4-heptanone undergoes fragmentation via a McLafferty rearrangement.

    Answer:

    $H_2C - \overset{\overset{\displaystyle H \diagdown \overset{\bullet}{\underset{+}{O}}\bullet}{\|}}{C} - CH_2CH_2CH_3$

    Section: 5

70. Arrange the following regions of the electromagnetic spectrum in order of decreasing wavelength: microwaves, visible light, ultraviolet light, infrared radiation, X-rays.

    Answer:
    microwaves > infrared radiation > visible light > ultraviolet light > X-rays
    Section: 6

MULTIPLE CHOICE

71. Which of the following is highest in energy per photon?

    a. Electromagnetic radiation with a wavenumber of 10^{-2} cm^{-1}.
    b. Electromagnetic radiation with a frequency of 10^7 cm/s.
    c. Electromagnetic radiation with a frequency of 10^5 cm/s.
    d. Electromagnetic radiation with a wavelength of 10 cm.
    e. Electromagnetic radiation with a wavenumber of 10^6 cm.

Answer: d   Section: 6

72. Which of the following functional groups typically exhibits a carbonyl stretch at the lowest frequency?

    a. amide
    b. ester
    c. ketone
    d. aldehyde
    e. ether

Answer: a   Section: 10

SHORT ANSWER

73. Explain how hydrogen bonding affects the shape of O-H and N-H absorption bands in IR spectra.

Answer:
O-H and N-H bands are generally quite broad. These bands are broad because the hydrogen bonds vary in strength. The more concentrated the sample, the more likely intermolecular hydrogen bonding can occur.
Section: 10

74. While cleaning the organic stockroom, a student found a bottle of ethanol labeled "Denatured with Benzene." He decided to determine the concentration of benzene in the ethanol by acquiring the UV spectrum of the liquid in a 2.0 cm cell. The spectrum exhibited an absorption band at 260 nm which was attributable to benzene, and the absorbance of this band was 0.69. In a reference source, the student found that the molar absorptivity of this benzene band in ethanol was 230 M^{-1}cm^{-1}. Show how the student calculated the concentration of the benzene in the ethanol.

Answer:
$A = cl\varepsilon$ $c = A/l\varepsilon$
                       $c = (0.69)/(230 \text{ M}^{-1}\text{cm}^{-1})(2.0 \text{ cm})$
                       $c = 0.0015 \text{ M}$

Section: 17

75. Is the λ_{max} of phenol in aqueous solution affected by the pH of the solution? If so, explain how and why λ_{max} changes.

Answer:
As the pH of the solution increases, an increasing amount of the phenol is converted into the conjugate base, the phenolate ion. The λ_{max} of the phenolate ion is greater than that of phenol since an additional pair of nonbonding electrons is interacting with the aromatic ring. This is known as a red shift.

Section: 18

Chapter 13: NMR Spectroscopy

MULTIPLE CHOICE

1. Which of the following spectroscopic techniques uses the lowest energy of the electromagnetic radiation spectrum?

   a. UV
   b. visible
   c. IR
   d. X-ray
   e. NMR

   Answer: e Section: 1

2. Which of the following is used in nuclear magnetic resonance spectroscopy?

   a. X-ray
   b. infrared
   c. visible
   d. radio
   e. ultraviolet

   Answer: d Section: 1

3. If the frequency for flipping an ^1H nucleus at an applied field of 1.4092 Tesla is 60 MHz, what would be the applied magnetic field if the frequency is 360 MHz?

   a. 8.4552 Tesla
   b. 0.2349 Tesla
   c. 1.4092 Tesla
   d. 4.2577 Tesla
   e. 3.0439 Tesla

   Answer: a Section: 1

4. How many signals would you expect to see in the ^1H NMR spectrum of the following compound?

$$CH_3CH_2CH_2\overset{\overset{\textstyle O}{\textstyle \|}}{C}CH_3$$

   a. 2
   b. 4
   c. 3
   d. 5
   e. 6

   Answer: b Section: 4

5. How many signals would you expect to see in the ^{1}H NMR spectrum of the following compound?

$$CH_3CH_2CH_2CH_3$$

   a. 1
   b. 3
   c. 2
   d. 4
   e. 6

Answer: c Section: 4

6. How many signals would you expect to see in the ^{1}H NMR spectrum of the following compound?

   a. 6
   b. 3
   c. 5
   d. 4
   e. 2

Answer: d Section: 4

7. How many signals would you expect to see in the ^{1}H NMR spectrum of the following compound?

   a. 6
   b. 5
   c. 4
   d. 3
   e. 2

Answer: e Section: 4

8. How many signals would you expect to see in the ¹H NMR spectrum of the following compound?

   a. 1
   b. 2
   c. 3
   d. 4
   e. 5

Answer: c Section: 4

9. How many signals would you expect to see in the ¹H NMR spectrum of the following compound?

   a. 5
   b. 4
   c. 2
   d. 3
   e. 1

Answer: d Section: 4

10. How many signals would you expect to see in the ¹H NMR spectrum of the following compound?

   a. 2
   b. 5
   c. 1
   d. 4
   e. 3

Answer: a Section: 4

11. How many signals would you expect to see in the ^{1}H NMR spectrum of the following compound?

    a. 1
    b. 2
    c. 4
    d. 5
    e. 3

Answer: e Section: 4

12. How many signals would you expect to see in the ^{1}H NMR spectrum of the following compound?

    a. 7
    b. 3
    c. 4
    d. 2
    e. 8

Answer: b Section: 4

13. How many signals would you expect to see in the ^{1}H NMR spectrum of the following compound?

$$CH_3CH_2CHCH_3$$
$$|$$
$$Cl$$

    a. 1
    b. 2
    c. 4
    d. 3
    e. 6

Answer: c Section: 4

14. How many signals would you expect to see in the ^{1}H NMR spectrum of the following compound?

$$ClCH_2CH_2Cl$$

   a. 5
   b. 4
   c. 3
   d. 2
   e. 1

Answer: e Section: 4

15. If a chemical shift of an NMR signal is 7.2 ppm measured in a 60 MHz NMR spectrometer, how many Hz would this signal be from the TMS signal?

   a. 8.3 Hz
   b. 432 Hz
   c. 0.12 Hz
   d. 72 Hz
   e. 60 Hz

Answer: b Section: 5

16. If two signals differ by 3 ppm, how do they differ in Hertz in a 60 MHz spectrometer?

   a. 20 Hz
   b. 300 Hz
   c. 360 Hz
   d. 180 Hz
   e. 90 Hz

Answer: d Section: 5

17. What is the ratio of the protons in the following compound?

   a. 3:3:3:2
   b. 6:3:2
   c. 9:2
   d. 3:2
   e. 6:2

Answer: c Section: 8

18. What is the ratio of the protons in the following compound?

a. 3:3:2
b. 3:2
c. 6:2:1
d. 3:1
e. 3:2:1

Answer: d Section: 8

19. Which of the following protons gives an NMR signal with the highest chemical shift value (furthest downfield)?

$$F—\underset{1}{CH_2}\underset{2}{CH_2}\underset{3}{CH_2}\underset{4}{CH_2}\underset{5}{CH_2}—Br$$

a. 1
b. 2
c. 3
d. 4
e. 5

Answer: a Section: 6

20. Which of the following protons gives an NMR signal with the lowest chemical shift value (furthest upfield)?

$$F—\underset{1}{CH_2}\underset{2}{CH_2}\underset{3}{CH_2}\underset{4}{CH_2}\underset{5}{CH_2}—Br$$

a. 1
b. 2
c. 3
d. 4
e. 5

Answer: c Section: 6

21. Which of the following protons gives an NMR signal with the highest chemical shift value (furthest downfield)?

$$(CH_3)_2CH \!\!-\!\! O \!\!-\!\! CH_2CH_2CH_3$$
$$1 \quad\quad 2 \quad\quad\quad 3 \;\; 4 \;\; 5$$

a. 1
b. 2
c. 3
d. 4
e. 5

Answer: b Section: 6

22. Which of the following compounds gives the highest chemical shift value (furthest downfield) in the NMR spectrum?

a. I
b. II
c. III
d. IV
e. V

   I. $(CH_3)_4C$

   II. $(CH_3)_4Si$

   III. $(CH_3)_3N\colon$

   IV.

   V.

Answer: e Section: 7

23. Which of the following protons gives an NMR signal with the lowest chemical shift value (furthest upfield)?

a. 1
b. 2
c. 3
d. 4
e. 5

$$\underset{1}{CH_3}\underset{2}{CH_2}O\underset{3}{CH_2}\!\!-\!\!\underset{4}{\bigcirc}\!\!-\!\!\underset{5}{CH_3}$$

Answer: a Section: 7

24. What splitting pattern is observed in the proton NMR spectrum for the indicated hydrogens?

    a. Singlet
    b. Doublet
    c. Triplet
    d. Quartet
    e. Septet

Answer: d Section: 10

25. What splitting pattern is observed in the proton NMR spectrum for the indicated hydrogens?

    a. Singlet
    b. Doublet
    c. Triplet
    d. Quartet
    e. Septet

Answer: c Section: 10

26. What splitting pattern is observed in the proton NMR spectrum for the indicated hydrogens?

$$\overset{\displaystyle O}{\overset{\displaystyle \|}{CH_3OCCH_2CH_3}}$$
$$\uparrow$$

    a. Singlet
    b. Doublet
    c. Triplet
    d. Quartet
    e. Septet

Answer: a Section: 10

27. What splitting pattern is observed in the proton NMR spectrum for the indicated hydrogens?

$$CH_3OCH_2CH_2OCH_3$$
$$\uparrow$$

a. Singlet
b. Doublet
c. Triplet
d. Quartet
e. Septet

Answer: a   Section: 10

28. An unknown compound, C_9H_{12}, gave the following NMR spectrum:

Triplet at 1.21 ppm (3H)
Singlet at 2.30 ppm (3H)
Quartet at 2.60 ppm (2H)
Singlet at 7.04 ppm (4H)
What is the structure of the compound?

a. I
b. II
c. III
d. IV
e. V

I.   CH₂CH₂CH₃ on benzene ring
I. $CH_2CH_2CH_3$

II. CH_2CH_3 , CH_3

III. $CHCH_3$, CH_3

IV. CH_3 , CH_3 , CH_3

V. CH_3 , CH_3 , CH_3

Answer: b   Section: 11

29. Which of the following methyl groups will exhibit the most downfield (highest) chemical shift in ^{1}H NMR spectroscopy?

    a. I
    b. II
    c. III
    d. IV
    e. V

    I. CH_3NH_2
    II. CH_3OH
    III. CH_3CCH_3
            ‖
            O
    IV. $CH_3CH_2CH_3$
    V.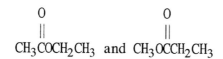

    Answer: b Section: 6

30. How many proton NMR singlets will 2-bromo-3-methyl-2-butene exhibit?

    a. 1
    b. 2
    c. 3
    d. 4
    e. 5

    Answer: c Section: 4

31. Which of the following technique(s) can readily distinguish between:

$$\underset{CH_3COCH_2CH_3}{\overset{O}{\overset{\|}{}}} \text{ and } \underset{CH_3OCCH_2CH_3}{\overset{O}{\overset{\|}{}}}$$

    a. NMR
    b. IR
    c. MS
    d. a and b
    e. a and c

    Answer: e Section: 11

SHORT ANSWER

32. Which of the following compounds has a signal disappear when D_2O is added to it, ethyl alcohol or ethyl chloride?

    Answer: The signal would disappear for ethyl alcohol and not for ethyl chloride.

    Section: 15

33. An unknown compound, $C_4H_8Br_2$, gave the following proton NMR data:
    Singlet at 1.97 ppm (6H)
    Singlet at 3.89 ppm (2H)
    What is the compound?

Answer:

singlet at 1.97 ppm {
$$CH_3 \quad Br$$
$$C$$
$$CH_3 \quad CH_2Br$$
singlet at 3.89 ppm

Section: 11

34. An unknown compound, $C_3H_5Cl_3$, gave the following proton NMR data:
    Doublet at 1.70 ppm (3H)
    Multiplet at 4.32 ppm (1H)
    Doublet at 5.85 ppm (1H)
    What is the structure of the compound?

Answer:

multiplet at 4.32 ppm

$$Cl$$
$$\quad CHCHCH_3 \longleftarrow \text{doublet at 1.70 ppm}$$
$$Cl \quad Cl$$
doublet at 5.85 ppm

Section: 11

35. An unknown compound, $C_4H_{10}O$, gave the following proton NMR data:
    Triplet at 1.13 ppm
    Quartet at 3.38 ppm
    What is the structure of the compound?

Answer:

quartet at 3.38 ppm

$$CH_3CH_2OCH_2CH_3$$

triplet at 1.13 ppm

Section: 11

36. Deduce the identity of the following compound from the spectral data given.

    $C_9H_{10}O_2$: ^{13}C NMR, δ 18.06 (quartet), 45.40 (doublet), 127.32 (doublet), 127.55 (doublet), 128.61 (doublet), 139.70 (singlet), 180.98 (singlet); IR, broad 3500-2800, 1708 cm^{-1}

    Answer: $PhCH(CH_3)CO_2H$ Section: 11

37. Deduce the identity of the following compound from the spectral data given.

    $C_5H_{10}O$: ^1H NMR, δ 1.2 (6H, doublet), 2.1 (3H, singlet), 2.8 (1H, septet); IR, 2980, 1710 cm^{-1}; MS, m/z 71, 43

    Answer: $(CH_3)_2CHCOCH_3$ Section: 11

38. Deduce the identity of the following compound from the spectral data given.

    $C_4H_8O_2$: ^1H NMR, δ 1.23 (3H, triplet), 2.00 (3H, singlet), 4.02 (2H, quartet); IR, 2980, 1740 cm^{-1}

    Answer: $CH_3CO_2CH_2CH_3$ Section: 11

39. Deduce the identity of the following compound from the spectral data given.

    $C_7H_{10}O_2$: ^1H NMR, δ 1.16 (3H, singlet), 2.21 (2H, singlet); ^{13}C NMR, δ 216.25 (singlet), 52.57 (singlet), 34.51 (triplet), 20.22 (quartet)

    Answer: 2,2-dimethyl-1, 3-cyclopentanedione Section: 11

40. Deduce the identity of the following compound from the ^1H NMR data given.

    C_4H_7BrO: δ 2.2 (3H, singlet), 3.5 (2H, triplet), 4.5 (2H, triplet)

    Answer: $CH_3COCH_2CH_2Br$ Section: 11

41. Deduce the identity of the following compound from the ^1H NMR data given.

    $C_8H_{10}O$: δ 3.4 (3H, singlet), 4.5 (2H, singlet), 7.2 (5H, singlet)

    Answer: $PhCH_2OCH_3$ Section: 11

42. Predict the number of signals expected, their splitting, and their relative area in the ^{1}H NMR spectrum of 1,2-dichloroethane (ClCH$_2$CH$_2$Cl).

Answer: 1 signal: singlet   Section: 11

43. Predict the number of signals expected, their splitting, and their relative area in the ^{1}H NMR spectrum of (CH$_3$)$_3$CCHO.

Answer: 2 signals: (9H, singlet); (1H, singlet)   Section: 11

44. Predict the number of signals expected, their splitting, and their relative area in the ^{1}H NMR spectrum of CH$_3$CH$_2$OCH$_3$.

Answer: 3 signals: (3H, triplet); (2H, quartet); (3H, singlet)

Section: 11

45. What multiplicities are observed in the spin coupled ^{13}C NMR spectrum of 2,3-dimethyl-2-butene?

Answer: a singlet and a quartet   Section: 18

46. How might the two trimethylcyclohexane isomers shown below be most readily distinguished using NMR?

Answer:

Number of signals in ^{13}C NMR, 3 signals versus 6.
Section: 18

47. Why is Fourier transform NMR spectroscopy preferred over continuous wave as a technique for ^{13}C NMR?

Answer:

^{13}C nuclei have a low sensitivity which requires multiple acquisitions of the spectrum before data become useful. The time/acquisition is much lower using FT techniques so that ^{13}C spectra can be acquired in reasonable times using this method.
Section: 2

48. Give one reason why ^{13}C NMR is less sensitive than ^1H NMR.

    Answer:

    Natural isotopic abundance of ^{13}C is about 100 times less than that of ^1H or gyromagnetic ratio of ^{13}C is much smaller.
    Section: 18

49. How might the proton spectrum of ultrapure dimethylamine, $(CH_3)_2NH$, differ from the spectrum of this compound to which D_2O has been added?

    Answer:
    N-H signal will broaden or disappear upon addition of D_2O as rapid hydrogen exchange occurs.
    Section: 15

50. The chair form of cyclohexane has protons in two distinct environments, axial and equatorial. When the proton NMR of cyclohexane is run on a 100 MHz instrument at 23° C, only one signal for the compound is observed. Explain this apparent contradiction.

    Answer:
    Axial and equatorial positions are being rapidly interchanged by chair-chair conformational interconversions. This rapid exchange results in an average signal for these two positions. At very low temperatures, two distinct signals are observed as the conformational interconversion process is slowed.
    Section: 14

MULTIPLE CHOICE

51. Using a 60 MHz spectrometer, the protons in dichloromethane appear at 5.30 ppm. When the same sample is placed in a 100 MHz instrument, where does the signal appear?

    a. 8.33
    b. 3.18
    c. 5.30
    d. cannot be determined from information given

    Answer: c Section: 5

SHORT ANSWER

52. _____ is commonly used an internal reference in NMR spectroscopy; its signal is assigned δ=0 in ^{1}H and ^{13}C NMR spectroscopy.

Answer: Tetramethylsilane, $(CH_3)_4Si$ Section: 5

MULTIPLE CHOICE

53. ^{1}H nuclei located near electronegative atoms tend to be _____ relative to ^{1}H nuclei which are not.

a. shielded
b. deshielded
c. resonanced
d. split
e. none of the above

Answer: b   Section: 3

54. How many signals will be observable in the DEPT 13C NMR spectrum of naphthalene (shown below)?

a. 0
b. 1
c. 2
d. 3
e. 10

Answer: c   Section: 19

SHORT ANSWER

55. What is indicated by a cross peak in the COSY spectrum of a compound?

Answer:
The dots that are not on the diagonal of a COSY spectrum are called cross peaks. Cross peaks indicate pairs of protons that are coupled to each other.
Section: 20

56. Deduce the identity of the compound from the data provided.

$C_5H_{10}O_2$: IR (cm^{-1}): 2950, 1740; ^{13}C NMR (δ, splitting): 15.8 (q), 19.7 (q), 68.4 (d), 195.3 (s)

Answer: $CH_3CO_2CH(CH_3)_2$ Section: 18

MULTIPLE CHOICE

57. Deduce the identity of the compound from the data provided.

C_7H_{12}: IR (cm^{-1}): 3300, 2950, 2220; ^{13}C NMR: 5 signals

a. 1,2-dimethylcyclopentene
b. 5-methyl-2-hexyne
c. 3,3-dimethyl-1-pentyne
d. 4,4-dimethyl-1-pentyne
e. 2,4-dimethyl-2-pentene

Answer: d   Section: 18

SHORT ANSWER

58. Deduce the identity of the compound from the data provided.

$C_{10}H_{14}O$:   IR (cm^{-1}): 3200-3500 (broad), 3050, 2950, 1610
$_1$H NMR(δ): 1.0 (s, 6H), 2.0 (s, 3H), 2.8 (broad s, 1H), 7.3 (d, 2H), 7.6 (d, 2H)

Answer:

Section: 11

59. Why is carbon-hydrogen coupling not generally seen in ^{1}H NMR spectra?

Answer:

Most carbons are ^{12}C. This nucleus has only one nuclear spin state and thus does not split attached H's in the ^{1}H NMR. When bound to a ^{13}C, a 1H nucleus does show coupling, but since only about 1% of carbon is ^{13}C, such coupling is small in magnitude.

Section: 18

MULTIPLE CHOICE

60. Deduce the identity of the compound from the data provided.

$C_5H_8O_4$:   IR (cm^{-1}): 2800-3300 (broad), 2950, 1740
^{13}C NMR (δ, splitting): 17.3 (q), 44.3 (s), 210.5 (s)

a. $HO_2CC(CH_3)_2CO_2H$
b. $CH_3O_2CCH_2CO_2CH_3$
c. $HO_2CCH_2CH_2CH_2CO_2H$
d. $CH_3O_2CCH_2CH_2CO_2H$
e. $CH_3O_2CCO_2CH_2CH_3$

Answer: a   Section: 18

SHORT ANSWER

61. H-H coupling is observed in the 1H NMR spectrum of $CH_3CH_2OCH_3$ and the signal from the methylene H's does not appear as a single peak. How does this signal appear? Explain in detail what is occurring.

Answer:
The methylene H's are perturbed by the nuclear spins of the three neighboring H's of the methyl group. The nuclear spins of these three neighboring H's can generate 4 distinct local magnetic environments: (all three spin up; one way), (two spin up and one spin down; 3 ways), (one spin up and two spin down; 3 ways), (three spin down; 1 way). Thus, the signal from the methylene H's is split into a quartet with relative signal areas of 1:3:3:1.

Section: 10

MULTIPLE CHOICE

62. How many distinct triplets would you expect in the ^1H NMR of the compound below?

a. 0
b. 2
c. 4
d. 6
e. 8

Answer: b   Section: 10

63. Deduce the identity of the compound from the data provided.

$C_8H_{13}Br$:   IR (cm^{-1}): 2950, 2150
             1H NMR (δ, splitting, integral): 3.5 (t, 2H), 1.8 (t, 2H),
             0.9 (s, 9H)
             ^{13}C NMR: 6 signals

a. $CH_3CHBrCCC(CH_3)_3$
b. $HCCCH_2C(CH_3)_2CH_2CH_2Br$
c. 3-bromo-1,2-dimethylcyclohexene
d. 4-bromo-1,2,4-trimethylcyclopentene
e. $BrCH_2CH_2CCC(CH_3)_3$

Answer: e   Section: 18

Chapter 14:   Reactions of Benzene

MULTIPLE CHOICE

1. Which of the following is an aromatic hydrocarbon?

    a. I
    b. II
    c. III
    d. IV
    e. V

I.

II.

III. CH_3

IV.

V. $CH=CH_2$

Answer: c   Section: 2

2. Which of the following compounds is aromatic?

    a. I
    b. II
    c. III
    d. IV
    e. V

I.

II.

III.   CH₃

IV.

V.

Answer: d Section: 3

3. Which of the following is aromatic?

   a. I
   b. II
   c. III
   d. IV
   e. V

   I.

   II.

   III.

   IV.

   V.

   Answer: c Section: 2

4. Which of the following is aromatic?

   a. I
   b. II
   c. III
   d. IV
   e. V

   I.

   II.

   III.

   IV.

   V. H   H
      N
      ⊕

   Answer: a Section: 3

5. Which of the following is aromatic?

    a. I
    b. II
    c. III
    d. IV
    e. V

I.

II.

III.

IV.

V.

Answer: b Section: 2

6. Which of the following is the most acidic?

    a. I
    b. II
    c. III
    d. IV
    e. V

I. CH_3CH_2-H

II. $CH_2=CH-H$

III. $CH\equiv C-H$

IV.

V.

Answer: e Section: 4

7. Which of the following is <u>antiaromatic</u>?

    a. I
    b. II
    c. III
    d. IV
    e. V

    I.

    II.

    III.

    IV.

    V.

Answer: d Section: 5

8. Which species is represented by the following distribution of π electrons in the molecular energy diagram?

   a. I
   b. II
   c. III
   d. IV
   e. V

$$\underline{\quad}\,\Psi_2 \quad \underline{\quad}\,\Psi_3$$
$$\underline{\downarrow\uparrow}\,\Psi_1$$

I.

II.

III. $\oplus$

IV. $\ominus$

V.

Answer: c Section: 6

SHORT ANSWER

9. The following molecule is not aromatic. Why?

Answer:
It does not have an odd number of pairs of pi electrons (i.e., it does not obey Huckle's rule) and it is not planar.
Section: 1

MULTIPLE CHOICE

10. Which of the following is most likely to be the <u>first step</u> in the general mechanism for electrophilic substitution reactions?

    a. I
    b. II
    c. III
    d. IV
    e. V

Answer: b   Section: 8

11. Which of the following is the electrophile that attacks the aromatic ring during sulfonation?

    a. $HSO_3^{\oplus}$

    b. $SO_2^{\oplus}$

    c. $HSO_3^{\ominus}$
    d. H_2SO_4

    e. $HSO_4^{\ominus}$

Answer: a   Section: 11

307

12. Which of the following is the electrophile that attacks the aromatic ring during nitration?

    a. NO_2
    b. $H\tilde{N}O_3$
    c. $\overset{\ominus}{N}O_3$
    d. $\overset{\oplus}{N}O_2$
    e. $\overset{\ominus}{N}O_2$

Answer: d   Section: 10

13. Which of the following is the first step in the mechanism of bromination?

    a. I
    b. II
    c. III
    d. IV
    e. V

Answer: c   Section: 9

14. Which of the following is the electrophile that attacks the aromatic ring during Friedel-Crafts acylation?

    a.   O
$$\text{R}-\overset{\overset{\displaystyle \parallel}{}}{\text{C}}-\text{Cl}$$

    b. $\text{R}-\text{C}\!\!=\!\!\text{O}$
           $\oplus$

    c.   O
$$\text{R}-\overset{\overset{\displaystyle \parallel}{}}{\text{C}}-\overset{\oplus}{\text{AlCl}_3}$$

    d.   O $\ominus$ $\oplus$
$$\text{R}-\overset{\overset{\displaystyle \parallel}{}}{\text{C}}-\overset{\delta}{\text{Cl}}\,\text{---}\,\overset{\delta}{\text{Cl}}$$

    e. AlCl_3

Answer: b Section: 12

15. What is the major product of the following Friedel-Crafts alkylation?

    a. I
    b. II
    c. III
    d. IV
    e. V

I. $\text{C}_6\text{H}_5-\text{CH}_2\text{CH}_2\text{CH}_2\text{CH}_3$

II. $\text{C}_6\text{H}_5-\text{CH}_2\text{CHCH}_3$, with CH_3 on the CH

III. $\text{C}_6\text{H}_5-\text{C(CH}_3)_3$

IV. $\text{C}_6\text{H}_5-\text{CH}_2\text{CH(CH}_3)_2$

V. $\text{C}_6\text{H}_5-\text{CH(CH}_3)\text{CH}_2\text{CH}_3$

Answer: c Section: 13

16. Which of the following is the best method for accomplishing this reaction?

    a. I
    b. II
    c. III
    d. I and II
    e. II and III

I. $CH_3CH_2CH_2Cl/AlCl_3$

II. $CH_3CH_2\overset{O}{\overset{\|}{C}}Cl/AlCl_3$; $Zn(Hg)/HCl/heat$

III. $CH_3CH_2\overset{O}{\overset{\|}{C}}Cl/AlCl_3$; $NH_2NH_2/\overset{\ominus}{O}H$, heat

Answer: e Section: 14

SHORT ANSWER

17. One of the following compounds is aromatic and the other is antiaromatic. Which is which?

Answer: a) is aromatic, b) is antiaromatic. Section: 3

18. Is the following molecule aromatic? Explain.

Answer:
Yes, it is aromatic since the molecule is planar, monocyclic with 5 pairs of π electrons.

Section: 1

19. Is the following molecule aromatic? Explain.

Answer: No, it is not aromatic since the molecule is not a conjugated system.

Section: 1

20. Draw the distribution of π electrons in the molecular orbitals of cyclobutadiene.

Answer:

antibonding m
nonbonding m
bonding mo

Section: 6

21. Cyclic hydrocarbons which can be represented as structures containing alternating single and double bonds are called _____.

Answer: annulenes   Section: 2

22. Provide the major resonance structures of the intermediate sigma complex in the reaction of benzene with the generic electrophile E^+.

Answer:

Section: 8

23. List the criteria which compounds must meet in order to be considered aromatic.

Answer:
1. The structure must be cyclic.
2. Each atom in the ring must have an unhybridized p orbital.
3. The structure must be planar or nearly planar so that overlap of these p orbitals is effective.
4. The π network must contain 4n + 2 electrons (where n is a whole number), so that delocalization of the π electrons results in a lowering of the molecule's electronic energy.

Section: 1

MULTIPLE CHOICE

24. Aromatic molecules contain _____ π electrons.

a. no
b. 4n + 2 (with n an integer)
c. 4n + 2 (where n = 0.5)
d. 4n (with n an integer)
e. unpaired

Answer: b   Section: 1

SHORT ANSWER

25. Provide the major resonance structures of the ion which results when the most acidic hydrogen of cyclopentadiene is lost.

Answer:

Section: 4

26. When cycloheptatriene is deprotonated, an anion with seven resonance forms of equal energy can be drawn. Given this fact, explain why cycloheptatriene is only slightly more acidic than propene.

Answer: Each resonance form of the cycloheptatrienyl anion is antiaromatic.

Section: 5

312

27. Classify cyclopentadienyl cation as aromatic, antiaromatic, or nonaromatic. Assume planarity of the π network.

    Answer: antiaromatic Section: 5

28. Classify cyclopheptatrienyl cation as aromatic, antiaromatic, or nonaromatic. Assume planarity of the π network.

    Answer: aromatic Section: 5

29. Classify the compound below as aromatic, antiaromatic, or nonaromatic. Assume planarity of the π netowrk.

    Answer: aromatic Section: 5

30. Classify the compound below as aromatic, antiaromatic, or nonaromatic. Assume planarity of the π netowrk.

    Answer: aromatic Section: 5

31. Is cyclooctatetraene a planar molecule? Explain.

    Answer:
    No. In addition to increasing other forms of strain, a planar conformation would make this molecule antiaromatic.
    Section: 5

MULTIPLE CHOICE

32. Which of the structures below would be aromatic?

    a. I and IV
    b. I, III, and IV
    c. III and IV
    d. II

Answer: c Section: 2

SHORT ANSWER

33. Indicate whether the following molecule is aromatic, antiaromatic, or nonaromatic.

Answer: aromatic Section: 5

34. In the bromination of benzene using Br_2 and $FeBr_3$, is the intermediate carbocation aromatic? Explain.

    Answer:
    No, the carbocation is not aromatic since the ring contains an sp^3 center.

    Section: 8

35. Provide a detailed, stepwise mechanism for the reaction of benzene with Br_2 and $FeBr_3$. Make sure to include the activating reaction between Br_2 and $FeBr_3$ in your mechanism.

Answer:

Section: 9

36. Name the major organic product which results when 4-ethylbenzenesulfonic acid is heated in aqueous acid.

Answer: ethylbenzene Section: 11

37. Provide the structure of the major organic product of the following reaction.

Answer:

Section: 13

315

MULTIPLE CHOICE

38. Which of the following is aromatic?

    a. cyclopentadienyl cation
    b. 1,3-cyclohexadiene
    c. cyclobutenyl anion
    d. 1,3,5-hexatriene
    e. cycloheptatrienyl cation

    Answer: e   Section: 2

SHORT ANSWER

39. Provide the structure of the biologically significant heterocycle purine.

    Answer:

    Section: 3

40. Provide the structure of the biologically significant heterocycle pyrimidine.

    Answer:

    Section: 3

41. Does cycloheptatriene have a higher or lower pKa than cyclopenatdiene? Explain.

    Answer:
    Cycloheptatriene is a weaker acid than cyclopentadiene and thus has a higher pKa. Cyclopentadiene is more acidic because the cyclopentadienyl anion which results upon deprotonation is aromatic and thus more stable than the corresponding conjugate base of cycloheptatriene.
    Section: 4

MULTIPLE CHOICE

42. Which of the following statements about the π molecular orbital description of cyclobutadiene is <u>not</u> correct?

    a. Cyclobutadiene has two degenerate nonbonding π molecular orbitals.
    b. Cyclobutadiene has a single bonding π molecular orbital.
    c. Cyclobutadiene has two electrons in nonbonding π molecular orbitals.
    d. Cyclobutadiene has one electron in an antibonding π molecular orbital which makes it antiaromatic.
    e. None of the above.

    Answer: d   Section: 6

SHORT ANSWER

43. Why does benzene undergo electrophilic substitution rather than electrophilic addition?

    Answer:
    When benzene undergoes electrophilic substitution, the product retains the stability associated with an aromatic system. This stability is lost in electrophilic addition because the product is not aromatic.
    Section: 7

44. Which step in the general mechanism for electrophilic aromatic substitution is rate-determining? Explain.

    Answer:
    The first step involves additon of the electrophile to benzene to yield a nonaromatic cationic intermediate. The second step involves the loss of H+ from this cationic intermediate to yield an aromatic product. The rate-determining or slow step is the one in which the aromatic reagent is converted into a much less stable nonaromatic cation.
    Section: 8

45. What is the principle of microscopic reversibility?

    Answer:
    This principle states that the mechanism of a reaction in the reverse direction must retrace each step of the mechanism in the forward direction in microscopic detail. Both the forward and reverse reactions must have the same intermediates and transition states.
    Section: 11

46. Why must Friedel-Crafts acylation reactions be run with more than a single equivalent of $AlCl_3$?

Answer:
These reactions must be run with more than one equivalent of $AlCl_3$ because the product of the reaction contains a carbonyl group that can complex with $AlCl_3$.

Section: 12

MULTIPLE CHOICE

47. What is the major organic product of the reaction between benzene and isobutyl chloride in the presence of $AlCl_3$?

    a. *tert*-butylbenzene
    b. isobutylbenzene
    c. n-butylbenzene
    d. chlorobenzene
    e. *sec*-butylbenzene

    Answer: a   Section: 13

SHORT ANSWER

48. Bromination of benzene requires a Lewis acid catalyst while bromination of cyclohexene does not. Explain this difference in reactivity.

    Answer:
    An alkene is a better nucleophile than benzene, and thus the Br-Br bond does not have to be weakened to form a better electrophile before reaction.

    Section: 9

MULTIPLE CHOICE

49. What purpose does $FeCl_3$ serve in the electrophilic aromatic substitution reaction between chlorine and benzene?

    a. It serves as a radical initiator to produce the chlorine radical needed to propagate the chain reaction.
    b. It functions by destabilizing the carbocationic intermediate and thereby increases the rate of H^+ loss.
    c. It serves as a Lewis base catalyst by reacting with Cl_2 to generate chloride ions.
    d. It functions by destabilizing the benzene through formation of a π-complex.
    e. It serves as a Lewis acid catalyst by reacting with the Cl2 and thereby activates it toward attack by benzene's π electrons.

    Answer: e   Section: 9

50. Predict the major organic product of the following reaction.

    a. I
    b. II
    c. III
    d. IV
    e. V

I.

II.

III.

IV.

V.

    Answer: e   Section: 11

51. Which of the following is one of the resonance contributors of the intermediate produced in a Friedel-Crafts alkylation of benzene?

    a. I
    b. II
    c. III
    d. IV
    e. V

I.

II.

III.

IV.

V.

Answer: a Section: 8

Chapter 15:   Reactions of Substituted Benzenes

MULTIPLE CHOICE

1. What is the structure of styrene?

   a. I
   b. II
   c. III
   d. IV
   e. V

   I. benzene ring with CH=CH$_2$

   II. benzene ring with NH$_2$

   III. benzene ring with OCH$_3$

   IV. benzene ring with C≡N

   V. benzene ring with CH$_3$

   Answer: a   Section: 1

2. What is the structure of phenol?

 a. I
 b. II
 c. III
 d. IV
 e. V

I.

CH$_2$OH

II.

$\overset{O}{\overset{\|}{CH}}$

III.

OCH$_3$

IV.

OH

V.

SO$_3$H

Answer: d Section: 1

3. What is the structure of 3-phenylpentane?

   a. I
   b. II
   c. III
   d. IV
   e. V

I. $CH_3CHCH_2CH_2CH_2CH_3$

II. $CH_3CH_2CHCH_2CH_3$

III. $CH_3CH_2CHCH_2CH_3$

IV. $CH_3CH_2CHCH_2CH_3$

V. $CH_3CH_2CHCH_2CH_3$

Answer: e Section: 1

4. What is the name of the following compound?

   a. p-Dichlorobenzene
   b. 1,4-Dichlorobenzene
   c. Phenyldichloride
   d. a and b
   e. b and c

Answer: d Section: 1

5. What is the name of the following compound?

    a. m-Bromomethylbenzene
    b. m-Bromotoluene
    c. 3-Bromotoluene
    d. a and b
    e. b and c

Answer: e Section: 1

6. What is the structure of p-toluidine?

    a. I
    b. II
    c. III
    d. IV
    e. V

    I.

    II.

    III.

    IV.

    V.

Answer: c Section: 1

7. What is the name of the following compound?

   a. o-nitro-m-bromotoluene
   b. 3-bromo-6-nitrotoluene
   c. m-bromo-o-nitrotoluene
   d. 5-bromo-2-nitrotoluene
   e. 2-nitro-5-bromotoluene

Answer: d Section: 1

8. The name 2,4,6-tribromobenzene is incorrect. Which of the following is the correct name?

   a. tribromobenzene
   b. m,m-dibromobromobenzene
   c. 3,5-dibromobromobenzene
   d. 1,3,5-tribromobenzene
   e. m,m,m-tribromobenzene

Answer: d Section: 1

9. What is the <u>major</u> product of the following reaction?

    a. I
    b. II
    c. III
    d. IV
    e. V

I.

II.

III.

IV.

V.

Answer: d Section: 2

10. What is the <u>major</u> product of the following reaction?

    a. I
    b. II
    c. III
    d. IV
    e. V

I.

COOH
COOH

II.

COOH
COOH

III.

COOH
CH₂CH₃

IV.

COOH
COOH

V.

COOH

Answer: a Section: 2

11. What is the major product of the following reaction?

   a. I
   b. II
   c. III
   d. IV
   e. V

I. Ph–C≡CCH₂CH₃   Ph(CH=CHCH₂CH₃) →[H₂ / Ni] ?

II. (cyclohexadiene)–CH=CHCH₂CH₃

III. (cyclohexene)–CH=CHCH₂CH₃

IV. (cyclohexane)–CH=CHCH₂CH₃

V. Ph–CH₂CH₂CH₂CH₃

Answer: e   Section: 2

12. What is the best method for carrying out the following reaction?

    a. I
    b. II
    c. III
    d. IV
    e. V

    I. $CH_3CH_2OH/AlCl_3$

    II. $CH_3CH_2Cl/AlCl_3$; NBS/heat ; $\overset{\ominus}{O}H$

    III. $CH_3CH_2Cl/AlCl_3$; $\overset{\ominus}{O}H$

    IV. $CH_3\overset{Cl}{\overset{|}{C}}HCl/AlCl_3$; $\overset{\ominus}{O}H$

    V. $CH_3\overset{O}{\overset{||}{C}}Cl/AlCl_3$; $KMnO_4$, heat

Answer: b Section: 8

13. Which of the following compounds reacts most <u>slowly</u> during nitration?

    a. I
    b. II
    c. III
    d. IV
    e. V

Answer: c Section: 3

14. Which of the following compounds reacts most rapidly during nitration?

   a. I
   b. II
   c. III
   d. IV
   e. V

I.   C≡N (benzene ring)

II.   NO$_2$ (benzene ring)

III.   CF$_3$ (benzene ring)

IV.   F (benzene ring)

V.   COOH (benzene ring)

Answer: d Section: 3

15. Which of the following substrates is an electron donating group overall?

   a. —Br
   b. —CCH$_3$
       ‖
       O
   c. —OCH$_3$
   d. $\oplus$
     —N(CH$_3$)$_3$
   e. —CCl$_3$

Answer: c Section: 3

16. Which of the following substrates is an electron withdrawing group overall?

    a. I
    b. II
    c. III
    d. IV
    e. V

    I. —CH_3

    II. —OH

    III. —

    IV. —OCH_3

    V. —SO_3H

    Answer: e   Section: 3

17. Which of the following structures is the most important contributor to the resonance hybrid formed when toluene undergoes para nitration?

a. I
b. II
c. III
d. IV
e. V

I.

II.

III.

IV.

V.

Answer: a   Section: 4

18. Which of the following structures is the most important contributor to the resonance hybrid formed when anisole undergoes o-bromination?

    a. I
    b. II
    c. III
    d. IV
    e. V

I.

II.

III.

IV.

V.

Answer: d Section: 4

19. What is/are the product(s) from the following reaction?

    a. I
    b. II
    c. III
    d. IV
    e. V

I.

II.

III.   CH₃NO₂

IV.

Answer: e   Section: 7

20. Which is the best method for carrying out the following reaction?

    a. ^+COOH; HNO_3/H_2SO_4
    b. $CH_3Cl/AlCl_3$; HNO_3/H_2SO_4; $KMnO_4/H^+$, heat
    c. $CH_3Cl/AlCl_3$; $KMnO_4/H^+$, heat; HNO_3/H_2SO_4
    d. HNO_3/H_2SO_4; $CH_3Cl/AlCl_3$; $KMnO_4/H^+$, heat
    e. HNO_3/H_2SO_4; ^+COOH

Answer: c   Section: 8

21. Which is the best method for carrying out the following reaction?

   a. I
   b. II
   c. III
   d. IV
   e. V

   I. SO_3/H_2SO_4 ; $CH_3CH_2CH_2Cl/AlCl_3$

   II. $CH_3CH_2CH_2Cl/AlCl_3$; SO_3/H_2SO_4

   III. $CH_3CH_2\overset{O}{\underset{||}{C}}Cl/AlCl_3$; SO_3/H_2SO_4

   IV. $CH_3CH_2\overset{O}{\underset{||}{C}}Cl/AlCl_3$; $Zn(Hg)/HCl/heat$; SO_3/H_2SO_4

   V. $CH_3CH_2\overset{O}{\underset{||}{C}}Cl/AlCl_3$; SO_3/H_2SO_4 ; $Zn(Hg)/HCl/heat$

Answer: e Section: 8

22. What is the <u>major</u> product of the following reaction?

    a. I
    b. II
    c. III
    d. IV
    e. V

    I. (benzene)—CH$_2$CH$_2$Br

    (benzene)—CH$_2$CH$_3$ + Br$_2$ $\xrightarrow{\text{FeBr}_3}$?

    II. (benzene)—CHCH$_3$ with Br

    III. (benzene, Br para)—CH$_2$CH$_3$

    IV. (benzene, two Br)—CH$_2$CH$_3$, Br

    V. (benzene, Br meta)—CH$_2$CH$_3$

    Answer: c Section: 4

23. Which of the following is the best method for preparing m-chloroaniline?

    a. NH$_3$; Cl$_2$/AlCl$_3$
    b. Cl$_2$/AlCl$_3$; NH$_3$
    c. Cl$_2$/AlCl$_3$; HNO$_3$/H$_2$SO$_4$; Sn/HCl, HO$^-$
    d. HNO$_3$/H$_2$SO$_4$; Cl$_2$/AlCl$_3$; Sn/HCl, HO-
    e. HNO$_3$/H$_2$SO$_4$; Sn/HCl; HO$^-$; Cl$_2$/AlCl$_3$

    Answer: d Section: 8

24. What is the <u>major</u> product of the following reaction?

a. I
b. II
c. III
d. IV
e. V

Answer: e Section: 9

25. What is the <u>major</u> product of the following reaction?

    a. I
    b. II
    c. III
    d. IV
    e. V

I.

II.

III.

IV.

V.

Answer: b Section: 9

26. What is the <u>best</u> method for the preparation of p-chlorotoluene in high yield?

    a. Start with benzene; methylate; chlorinate
    b. Start with benzene; chlorinate; methylate
    c. Start with toluene; chlorinate
    d. Start with chlorobenzene; methylate
    e. Start with p-aminotoluene; $NaNO_2$/HCl, 0° C; CuCl

    Answer: e Section: 10

27. What is the <u>best</u> method for the preparation of m-dibromobenzene from benzene?

    a. Nitrate; Sn/HCl; $NaNO_2$/HCl, 0° C; brominate twice
    b. Nitrate; Sn/HCl; $NaNO_2$/HCl, 0° C; brominate twice; H_3PO_2
    c. Nitrate; Sn/HCl; $NaNO_2$/HCl, 0° C; H_3PO_2; brominate twice
    d. Nitrate; brominate; Sn/HCl; $NaNO_2$/HCl, 0° C; CuBr
    e. Brominate twice

    Answer: d Section: 10

28. What is one of the products of the following reaction?

   a. I
   b. II
   c. III
   d. IV
   e. V

Answer: c Section: 11

29. Which of the following is <u>most</u> reactive toward nucleophilic aromatic substitution?

   a. I
   b. II
   c. III
   d. IV
   e. V

Answer: d   Section: 13

30. Which of the following is known as benzyne?

    a. I
    b. II
    c. III
    d. IV
    e. V

    I.

    II.

    III.   C≡CH

    IV.

    V.

Answer: d Section: 14

31. What are the products of the following reaction?

    a. I
    b. II
    c. III
    d. a and b
    e. b and c

Answer: d Section: 14

32. What is the <u>major</u> product from the following reaction?

    a. I
    b. II
    c. III
    d. a and b equally
    e. a and c equally

$$\text{naphthalene} + \xrightarrow[\text{H}_2\text{SO}_4]{\text{HNO}_3} \text{?}$$

I. (naphthalene with NO_2)

II. (naphthalene with NO_2)

III. (naphthalene with NO_2)

Answer: a Section: 16

33. What is the <u>major</u> product of the following reaction?

    a. I
    b. II
    c. III
    d. a and b equally
    e. a and c equally

Answer: c Section: 16

SHORT ANSWER

34. What is the name of the following compound?

Answer: 1,3-Dinitronaphthalene Section: 16

35. Explain why nitrobenzene can be used as a solvent for Friedel-Crafts alkylation.

    Answer:
    Because of the deactivating nature of the NO_2 group, it will not undergo a F.C. reaction.
    Section: 7

36. Explain why direct nitration of aniline yields, among other products, m-nitroaniline.

    Answer:
    In acidic media, the basic amino group forms a protonated ammonium ion which is a meta director.
    Section: 7

37. Rank the following groups in order of increasing activating power in electrophilic aromatic substitution reactions: $-OCH_3$, $-OCOCH_2CH_3$, $-CH_2CH_3$, $-Br$.

    Answer: $-Br < -CH_2CH_3 < -OCOCH_2CH_3 < -OCH_3$ Section: 3

MULTIPLE CHOICE

38. In electrophilic aromatic substitution reactions a chlorine substituent:

    a. is a deactivator and a m-director.
    b. is a deactivator and an o,p-director.
    c. is an activator and a m-director.
    d. is an activator and an o,p-director.
    e. none of the above.

    Answer: b   Section: 4

SHORT ANSWER

39. Draw the four major resonance structures of the carbocation intermediate in the reaction of anisole with HNO_3/H_2SO_4 to yield p-nitroanisole.

    Answer:

    Section: 4

345

40. Draw the three major resonance structures of the carbocation intermediate in the reaction of acetophenone with HNO_3/H_2SO_4 to yield o-nitroacetophenone. Circle the resonance form which is less stable than the other two.

Answer:

Section: 4

41. Provide a series of synthetic steps by which 2-bromo-4-nitrobenzoic acid can be prepared from toluene.

Answer:
1. HNO_3, H_2SO_4
2. Br_2, $FeBr_3$
3. $KMnO_4$, H^+
Section: 9

42. Provide the structure of the major mononitration product of the compound below.

Answer:

Section: 9

43. Provide the structure of the major mononitration product(s) of the compound below.

Answer:

Section: 9

44. Provide the structure of the major organic product of the following reaction.

Answer:
No reaction. The ring is too deactivated by the acyl substituent to undergo Friedel-Crafts aklylation.

Section: 7

45. Provide a detailed, stepwise mechanism for the following reaction.

Answer:

Section: 13

46. Provide a detailed, stepwise mechanism for the following reaction.

Answer:

Section: 14

47. Draw the <u>four</u> major resonance structures of the intermediate which results when o-nitrochlorobenzene is treated with NaOH.

Answer:

Section: 13

48. Provide the structure of benzoic acid.

Answer:

Section: 1

49. Provide the structure of anisole.

Answer:

Section: 1

MULTIPLE CHOICE

50. What is the proper name of the compound below?

    a. 1-phenoxyethane
    b. ethyl phenyl ether
    c. m-butylhydroxybenzene
    d. o-butylhydroxybenzene
    e. m-butylphenol

    Answer: e   Section: 1

51. How many signals does one expect in the proton decoupled ^{13}C NMR spectrum of o-xylene?

    a. 8
    b. 6
    c. 4
    d. 3
    e. 2

    Answer: c   Section: 1

SHORT ANSWER

52. Can one distinguish among the three isomeric xylenes using ^{13}C NMR? Explain.

    Answer:

    Yes. In the proton decoupled ^{13}C NMR spectra of these three isomers, different numbers of peaks are present. The *ortho* isomer has 4, the *meta* isomer has 5, and the *para* iosmer has 3.

    Section: 1

MULTIPLE CHOICE

53. Which of the following compounds reacts most rapidly with HNO_3/H_2SO_4?

    a. toluene
    b. anisole
    c. nitrobenzene
    d. benzonitrile
    e. fluorobenzene

    Answer: b   Section: 3

54. Benzene can be reduced to cyclohexane when it is treated with hydrogen in the presence of a nickel catalyst under conditions of high temperature and pressure. Predict the signs of $\Delta H°$ and $\Delta S°$ for this process.

    a. $\Delta H° > 0$, $\Delta S° > 0$
    b. $\Delta H° > 0$, $\Delta S° < 0$
    c. $\Delta H° > 0$, $\Delta S° = 0$
    d. $\Delta H° < 0$, $\Delta S° > 0$
    e. $\Delta H° < 0$, $\Delta S° < 0$

    Answer: e   Section: 2

55. Which of the following fails to produce benzoic acid when heated in the presence of acidic sodium dichromate?

    a. 1-phenylethanol
    b. 2-phenylethanol
    c. 2-phenylheptane
    d. 2-methyl-2-phenylheptane
    e. 2-phenylethanol

    Answer: d   Section: 2

SHORT ANSWER

56. List the following compounds in order of decreasing reactivity toward electrophilic aromatic substitution: toluene, benzene, fluorobenzene, nitrobenzene, phenol.

    Answer: phenol > toluene > benzene > fluorobenzene > nitrobenzene

    Section: 3

57. List the following compounds in order of decreasing basicity.

Answer:

Section: 5

58. List the following compounds in order of decreasing acidity.

Answer:

Section: 5

59. Provide a series of synthetic steps by which p-bromoanisole can be prepared from benzene.

Answer:
1. Cl_2, $FeCl_3$
2. HNO_3, H_2SO_4
3. $NaOCH_3$
4. Sn, HCl
5. $NaNO_2$, HCl, $0°C$
6. $CuBr$

Section: 13

60. What factors affect the ortho/para ratio in electrophilic aromatic substitution reactions?

Answer:
Based solely on probability, one would expect more of the ortho product because there are two ortho positions available to the incoming electrophile versus only on para position. However, the ortho position is sterically hindered relative to the para position. Consequently, the para product will be generated preferentially if either the substituent on the ring or the attacking electrophile is large.

Section: 6

Chapter 16: Carbonyl Compounds I

MULTIPLE CHOICE

1. Which of the following compounds is an amide?

   a. I
   b. II
   c. III
   d. IV
   e. V

    I. $CH_3CH_2NH_2$

    II. CH_3NHNH_2

    III. $CH_3\underset{\underset{O}{\|}}{C}NH_2$

    IV. $HO\underset{\underset{O}{\|}}{C}CH_2NH_2$

    V. (benzene ring with NH_2 substituent)

Answer: c   Section: 1

2. Which of the following compounds is an acyl chloride?

   a. I
   b. II
   c. III
   d. IV
   e. V

I.

II.

III.

IV.

V.

Answer: b Section: 1

3. Which of the following compounds is an ester?

   a. I
   b. II
   c. III
   d. IV
   e. V

I. CH_3CO—⬡

II. CH_3C—⬡

III. CH_3COH

IV. CH_3COOH

V. CH_3COCCH_3

Answer: a Section: 1

4. Which of the following compounds is an anhydride?

   a. I
   b. II
   c. III
   d. IV
   e. V

I. $CH_3\overset{\overset{O}{\|}}{C}O\!-\!\bigcirc$

II. $CH_3\overset{\overset{O}{\|}}{C}\!-\!\bigcirc$

III. $CH_3\overset{\overset{O}{\|}}{C}OH$

IV. $CH_3\overset{\overset{O}{\|}}{C}OOH$

V. $CH_3\overset{\overset{O}{\|}}{C}O\overset{\overset{O}{\|}}{C}CH_3$

Answer: e   Section: 1

5. What is the common name for the following compound?

$$CH_3\underset{\underset{Cl}{|}}{C}HCH_2\overset{\overset{O}{\|}}{C}OH$$

   a. γ-Chlorobutanoic acid
   b. β-Chlorobutanoic acid
   c. β-Chlorobutyric acid
   d. γ-Chlorobutyric acid
   e. 3-Chlorobutyric acid

Answer: c   Section: 1

6. What is the IUPAC name for the following compound?

$$CH_3CH_2\underset{\underset{CH_3}{|}}{C}HCH_2COOH$$

   a. 3-Methylpentanoic acid
   b. Isohexanoic acid
   c. β-Methylvaleric acid
   d. 2-Methylpentanoic acid
   e. 3-Methylvaleric acid

Answer: a   Section: 1

7. Which of the following compounds is benzoyl chloride?

   a. I
   b. II
   c. III
   d. IV
   e. V

I. — benzene ring with CH₂Cl

II. — benzene ring with Cl

III. — benzene ring with OCH₂Cl

IV. — benzene ring with COCl

V. — benzene ring with COOCl

Answer: d Section: 1

8. What is the common name for the following compound?

$$\underset{\text{HC}}{\overset{\text{O}}{\|}} - \text{O} - \underset{\text{CH}}{\overset{\text{O}}{\|}}$$

   a. Methanoic anhydride
   b. Formic anhydride
   c. Formyl formate
   d. Ethanoic anhydride
   e. Diformyl ether

Answer: b Section: 1

9. What is the IUPAC name for the following compound?

    a. Phenyl acetate
    b. Methyl benzoate
    c. Methyl phenylacetate
    d. Benzyl acetate
    e. Benzyl methylacetate

Answer: d Section: 1

10. Which of the following compounds is methyl formate?

    a. $\overset{\displaystyle O}{\overset{\|}{CH_3CH}}$

    b. $\overset{\displaystyle O}{\overset{\|}{CH_3COCH_3}}$

    c. $\overset{\displaystyle O}{\overset{\|}{CH_3COH}}$

    d. $\overset{\displaystyle O}{\overset{\|}{HCOOCH_3}}$

    e. $\overset{\displaystyle O}{\overset{\|}{CH_3OCH}}$

Answer: e Section: 1

11. Which of the following compounds is γ-butyrolactone?

    a. I
    b. II
    c. III
    d. IV
    e. V

Answer: c Section: 1

12. What is the common name for the following compound?

    a. Phenylcarbamide
    b. Benzamide
    c. Phenylmethanamide
    d. Benzeneamide
    e. Benzylamide

Answer: b Section: 1

13. Which of the following compounds is N,N-dimethylbenzamide?

    a. I
    b. II
    c. III
    d. IV
    e. V

I.

II.

III.

IV.

V.

Answer: e Section: 1

14. Which of the following compounds is γ-valerolactam?

    a. I
    b. II
    c. III
    d. IV
    e. V

I.

II.

III.

IV.

V.

Answer: d Section: 1

15. What is the common name for the following compound?

    a. Benzonitrile
    b. Benzenecyanide
    c. Cyanobenzene
    d. Cyanophenyl
    e. Benzocyanide

$C\equiv N$

Answer: a Section: 1

16. What is the hybridization and geometry of the carbonyl carbon in carboxylic acids and their derivatives?

    a. sp^3, tetrahedral
    b. sp^2, trigonal planar
    c. sp^2, tetrahedral
    d. sp^3, trigonal planar
    e. sp, trigonal planar

    Answer: b Section: 2

17. Which of the following compounds is most reactive toward nucleophilic acyl substitution?

    a.
$$R-\overset{\displaystyle O}{\overset{\|}{C}}-NH_2$$
    b.
$$R-\overset{\displaystyle O}{\overset{\|}{C}}-OCH_3$$
    c.
$$R-\overset{\displaystyle O}{\overset{\|}{C}}-Cl$$
    d.
$$R-\overset{\displaystyle O}{\overset{\|}{C}}-O-\overset{\displaystyle O}{\overset{\|}{C}}-R$$
    e.
$$R-\overset{\displaystyle O}{\overset{\|}{C}}-OH$$

    Answer: c Section: 5

18. Which of the following compounds is the <u>least</u> reactive toward nucleophilic acyl substitution?

   a.

$$R-\overset{\displaystyle O}{\overset{\|}{C}}-NH_2$$

   b.

$$R-\overset{\displaystyle O}{\overset{\|}{C}}-OCH_3$$

   c.

$$R-\overset{\displaystyle O}{\overset{\|}{C}}-Cl$$

   d.

$$R-\overset{\displaystyle O}{\overset{\|}{C}}-O-\overset{\displaystyle O}{\overset{\|}{C}}-R$$

   e.

$$R-\overset{\displaystyle O}{\overset{\|}{C}}-OH$$

Answer: a Section: 5

19. Which of the following compounds is hydrolyzed <u>most</u> <u>rapidly</u> in aqueous NaOH?

   a.

$$R-\overset{\displaystyle O}{\overset{\|}{C}}-NH_2$$

   b.

$$R-\overset{\displaystyle O}{\overset{\|}{C}}-OCH_3$$

   c.

$$R-\overset{\displaystyle O}{\overset{\|}{C}}-Cl$$

   d.

$$R-\overset{\displaystyle O}{\overset{\|}{C}}-O-\overset{\displaystyle O}{\overset{\|}{C}}-R$$

   e.

$$R-\overset{\displaystyle O}{\overset{\|}{C}}-OH$$

Answer: c Section: 5

20. Which of the following compounds is hydrolyzed <u>most</u> <u>slowly</u> in aqueous NaOH?

a. $R-\overset{\overset{O}{\|}}{C}-O-\overset{\overset{O}{\|}}{C}-CH_3$

b. $R-\overset{\overset{O}{\|}}{C}-NH_2$

c. $R-\overset{\overset{O}{\|}}{C}-OCH_2-\bigcirc$

d. $R-\overset{\overset{O}{\|}}{C}-OCH_3$

e. $R-\overset{\overset{O}{\|}}{C}-Cl$

Answer: b Section: 5

21. Acetyl chloride undergoes nucleophilic substitution at a faster rate than methyl acetate because:

a. the ester is more sterically hindered than the acid chloride.
b. the acid chloride is more sterically hindered than the ester.
c. the methoxide is a better leaving group than chloride.
d. esters hydrolyze faster than acid chlorides.
e. chloride is a better leaving group than methoxide.

Answer: e Section: 5

22. Which of the following is the best method for preparing

$$CH_3-\overset{\overset{O}{\|}}{C}-O-CH_3?$$

a.
$$CH_3-\overset{\overset{O}{\|}}{C}-H + CH_3OH \xrightarrow{\overset{\oplus}{H}}$$

b.
$$CH_3-\overset{\overset{O}{\|}}{C}-Cl + CH_3\overset{\ominus}{O}\overset{\oplus}{Na} \longrightarrow$$

c.
$$CH_3-\overset{\overset{O}{\|}}{C}-H + CH_3\overset{\ominus}{O}\overset{\oplus}{Na} \longrightarrow$$

d.
$$CH_3-\overset{\overset{O}{\|}}{C}-OH + CH_3OH \xrightarrow{\overset{\oplus}{H}}$$

e.
$$CH_3-\overset{\overset{O}{\|}}{C}-NH_2 + CH_3\overset{\ominus}{O}\overset{\oplus}{Na} \longrightarrow$$

Answer: b Section: 7

23. Which of the following compounds would yield acetic acid when hydrolyzed under heat?

    a.
    $$\underset{CH_3C-O-CCH_3}{\overset{O\quad\quad O}{\quad||\quad\quad||\quad}}$$
    b. $CH_3C\equiv N$
    c.
    $$\underset{CH_3-C-Cl}{\overset{O}{||}}$$
    d.
    $$\underset{CH_3-C-OCH_3}{\overset{O}{||}}$$
    e. All the above

    Answer: e   Section: 7

24. Which of the following is the best method for preparing aspirin,

    a. I
    b. II
    c. III
    d. IV
    e. V

    Answer: e   Section: 8

25. Which of the following methods can be used to prepare

$$CH_3-C\equiv N?$$

a. $CH_3Br + {}^-C\equiv N \longrightarrow$

b.
$$\begin{array}{c} O \\ \parallel \\ CH_3CNH_2 \end{array} \xrightarrow[\text{heat}]{P_2O_5}$$

c. $CH_3CH_2Br + NH_2 \longrightarrow$

d. a and b

e. b and c

Answer: d Section: 13

26. Which of the following is one of the steps in the mechanism of the following reaction?

$$\overset{O}{\underset{\parallel}{R-C}}-OH + CH_3OH \xrightarrow{\overset{\oplus}{H}} \overset{O}{\underset{\parallel}{R-C}}-OCH_3 + H_2O$$

a.
$$\overset{O}{\underset{\parallel}{R-C}}-OH + CH_3OH \longrightarrow \overset{O^\ominus}{\underset{\underset{HOCH_3}{\mid}}{R-C}}-OH$$

b.
$$\overset{\overset{\oplus}{O}H}{\underset{\parallel}{R-C}}-OH + CH_3O^\ominus \longrightarrow \overset{\overset{\oplus}{O}H}{\underset{\underset{OCH_3}{\mid}}{R-C}}-OH$$

c.
$$\overset{\overset{\oplus}{O}H}{\underset{\parallel}{R-C}}-OH + CH_3OH \longrightarrow \overset{OH}{\underset{\underset{\overset{\mid}{H-O-CH_3}}{\mid}}{R-C}}-OH$$
$$\underset{\oplus}{}$$

d.
$$\overset{O}{\underset{\parallel}{R-C}}-OH + CH_3O^\ominus \longrightarrow \overset{O^\ominus}{\underset{\underset{OCH_3}{\mid}}{R-C}}-OH$$

e.
$$\overset{O}{\underset{\parallel}{R-C}}-OH + CH_3OH \longrightarrow \overset{O}{\underset{\underset{\oplus}{}}{R-C}} + {}^\ominus OH$$

Answer: c Section: 10

366

27. What is the product obtained from the following reaction?

    a. I
    b. II
    c. III
    d. IV
    e. V

I. benzoic acid methyl ester, $C-O-CH_3$, $+ H_2^{18}O$

II. $C-\overset{18}{O}-CH_3$, $+ H_2^{18}O$

III. $\overset{18}{C}-O-CH_3$, $+ H_2^{18}O$

IV. $\overset{18}{C}-O-CH_3$, $\overset{18}{O}$, $+ H_2O$

V. $\overset{18}{C}-O-CH_3$, $\overset{18}{O}$, $+ H_2O$

reaction: benzoic acid $C-OH$ $+ CH_3^{18}OH \xrightarrow{H^{\oplus}}$?

Answer: d   Section: 12

28. What is the <u>major</u> organic product obtained from the following sequence of reactions?

$$CH_3-\overset{O}{\overset{||}{C}}-OH \xrightarrow{SOCl_2} \xrightarrow{CH_3OH} ?$$

    a. $CH_3-\overset{O}{\overset{||}{C}}-O-CH_3$

    b. $\overset{O}{\overset{||}{CH_2}-C}-O-CH_3$ with Cl below CH_2

    c. $CH_3-O-\overset{O}{\overset{||}{C}}-H$

d.

$$CH_3-O-\overset{\overset{\displaystyle O}{\|}}{C}-H$$

e.

$$CH_3-\overset{\overset{\displaystyle O}{\|}}{C}-O-Cl$$

Answer: a Section: 18

29. Which of the following reagents is used in the following reaction?

a. Cl_2/P
b. Cl_2/CCl_4
c. $Cl_2/h\upsilon$
d. HCl
e. PCl_3

Answer: e Section: 18

30. Which of the following is phthalic acid?

   a. I
   b. II
   c. III
   d. IV
   e. V

I.

COOH

COOH

II.

COOH

HOOC

III.

COOH

COOH

IV.

COOH

COOH

COOH

V.

HOOC

COOH

COOH

Answer: c Section: 19

31. What is the <u>major</u> product of the following reaction?

    a. I
    b. II
    c. III
    d. IV
    e. V

I.

+ CO$_2$

$\xrightarrow{\text{heat}}$?

II.

+ H$_2$O

III.

+ CO$_2$

IV.

+ H$_2$O

V.

Answer: b Section: 19

32. What is the <u>major</u> organic product of the following reaction?

    a. I
    b. II
    c. III
    d. IV
    e. V

$$2 \text{ (benzene ring)} \text{COOH} + HOCH_2CH_2OH \xrightarrow{H^{\oplus}} ?$$

I. HOOC—(ring)—CH₂CH₂—(ring)—COOH

II. HOOC—(ring)—CH₂CH₂O—(ring)—COOH

III. (ring)—$\overset{O}{\overset{\|}{C}}$CH₂CH₂$\overset{O}{\overset{\|}{C}}$—(ring)

IV. (ring)—$\overset{O}{\overset{\|}{C}}$OCH₂CH₂O$\overset{O}{\overset{\|}{C}}$—(ring)

V. (ring)—$\overset{O}{\overset{\|}{C}}$$\overset{O}{\overset{\|}{C}}$—(ring)—OCH₂CH₂OH

Answer: d Section: 12

SHORT ANSWER

33. Do you expect the following reaction to go to completion? Why or why not?

$$R\!-\!\overset{O}{\overset{\|}{C}}\!-\!OCH_3 + NaCl \longrightarrow ?$$

Answer:
No. The equilibrium will favor the reactants because the nucleophile, Cl^-, is a much weaker base than the leaving group CH_3O^-.
Section: 4

34. What are the products from the following reaction?

Answer:

+ HCl

Section: 7

35. An unknown compound, $C_5H_{10}O_2$, gave the following proton NMR data:
   a) doublet, at 1.23 ppm (6H)
   b) singlet, at 2.10 ppm (3H)
   c) septet, at 4.98 ppm (1H)
      Propose a structure for the compound.

Answer:

36. What is the product of the following reaction?

Answer:

This is the hydrolysis of an anhydride to an acid.

Section: 8

37. Provide a detailed, stepwise mechanism for the reaction of acetyl chloride with methanol to produce methyl acetate and HCl.

Answer:

$$CH_3-\overset{\overset{O}{\|}}{C}-Cl + CH_3-OH \longrightarrow CH_3-\overset{\overset{O^-}{|}}{\underset{\underset{+}{CH_3-O-H}}{C}}-Cl \longrightarrow CH_3-\overset{\overset{O}{\|}}{C}-\overset{\overset{H}{|}}{\underset{+}{O}}CH_3 -$$

$$HCl + CH_3-\overset{\overset{O}{\|}}{C}-OCH_3 \longleftarrow$$

Cl⁻

Section: 7

38. What are the products of the reaction of benzoic acid with thionyl chloride?

Answer: benzoyl chloride, $SO_2(g)$, and HCl (g) Section: 18

39. Give a detailed, stepwise mechanism for the Fischer esterification of acetic acid with methanol.

Answer:

$$CH_3-\overset{\overset{O}{\|}}{C}-OH + H^+ \rightleftharpoons CH_3-\overset{\overset{\overset{+}{O}-H}{\|}}{C}-OH + CH_3OH \rightleftharpoons CH_3-\overset{\overset{O-H}{|}}{\underset{\underset{+}{CH_3-O-H}}{C}}-OH$$

$$\Big\downarrow -H^+$$

$$CH_3-\overset{+}{C}=O-H \quad + H_2O \qquad CH_3-\overset{\overset{\overset{H}{|}}{\overset{+}{O}-H}}{\underset{CH_3-O}{C}}-OH \overset{+H^+}{\rightleftharpoons} CH_3-\overset{\overset{O-H}{|}}{\underset{CH_3-O}{C}}-OH$$

$$H_2O + \qquad \overset{|}{OCH_3}$$

$$H_3O^+ + CH_3-\overset{\overset{O}{\|}}{C}-OCH_3$$

Section: 12

MULTIPLE CHOICE

40. Which of the following conditions will drive the equilibrium of the Fischer esterification towards ester formation?

   a. addition of water
   b. removal of water as it is formed
   c. addition of an inorganic acid as a catalyst
   d. addition of alcohol
   e. both b and d

   Answer: e   Section: 12

41. Esters and amides are most easily made by nucleophilic acyl substitution reactions on:

   a. alcohols
   b. acid anhydrides
   c. carboxylates
   d. carboxylic acids
   e. acid chlorides

   Answer: e   Section: 7

42. In nucleophilic acyl substitution,:

   a. protonation of the carbonyl is followed immediately by loss of the leaving group.
   b. loss of the leaving group is followed by rearrangement of the carbocation.
   c. addition to the carbonyl by a nucleophile is followed by loss of the leaving group.
   d. ester hydrolysis is followed by deprotonation.
   e. an S_N2 reaction occurs.

   Answer: c   Section: 4

SHORT ANSWER

43. Provide a detailed, stepwise mechanism for the reaction of propanoyl chloride with ammonia.

   Answer:

   Section: 7

44. Provide a detailed, stepwise mechanism for the reaction of propanoic anhydride with phenol.

Answer:

Section: 8

45. Provide a detailed, stepwise mechanism for the acid-catalyzed transesterification of ethyl acetate with n-propanol.

Answer:

Section: 10

46. Provide a detailed, stepwise mechanism for the base-mediated transesterification of methyl benzoate with sodium ethoxide.

Answer:

Section: 11

47. Provide a detailed, stepwise mechanism for the reaction of butyrolactone with ammonia.

Answer:

Section: 9

MULTIPLE CHOICE

48. The hydrolysis of esters in base is called:

    a. the Fischer esterification.
    b. the Hunsdiecker reaction.
    c. the Dieckmann condensation.
    d. transesterification.
    e. saponification.

    Answer: e Section: 17

49. Typically, amides will hydrolyze under _____ conditions than esters.

    a. milder
    b. more dilute
    c. stronger
    d. less vigorous
    e. more saline

    Answer: c Section: 5

50. Phthalic acid produces what acid derivative upon heating?

    a. an ester
    b. a carboxylate
    c. an acid chloride
    d. an anhydride
    e. an amide

    Answer: d Section: 19

SHORT ANSWER

51. What are the major organic products when 1 equivalent of acetic formic anhydride reacts with 1 equivalent of methanol?

    Answer: acetic acid and methyl formate Section: 8

MULTIPLE CHOICE

52. A nitrile can be made by dehydrating an amide. However, for this reaction to occur, the amide must be:

    a. primary.
    b. secondary.
    c. tertiary.
    d. N-methylated.
    e. part of a lactam.

    Answer: a Section: 13

SHORT ANSWER

53. What reagents are used to dehydrate amides to nitriles?

    Answer: P_2O_5 or $POCl_3$ Section: 13

54. Why are β-lactams, such as penicillins and cephalosporins, unusually reactive amides?

    Answer:
    Their enhanced reactivity is due to the considerable ring strain in the four-membered ring.
    Section: 14

55. Propose a synthesis of benzonitrile from benzene and any other necessary reagents.

    Answer:
    1. HNO_3, H_2SO_4
    2. Fe, HCl
    3. $NaNO_2$, HCl, cold
    4. CuCN
    Section: 18

56. Carbonated beverages typically have pH's below 7.0. What acid is present which lowers the pH? Show the reaction by which it is formed.

    Answer: carbonic acid, $CO_2 + H_2O \longrightarrow H_2CO_3$ Section: 19

57. Provide the proper IUPAC name for the compound below.

    Answer: isopropyl cyclohexanecarboxylate   Section: 1

58. Provide the proper IUPAC name for $ClCH_2CH_2CONHCH_3$.

    Answer: N-methyl 3-chloropropanamide   Section: 1

59. Provide the proper IUPAC name for the compound below.

    Answer: trans-5-octenoyl chloride   Section: 1

60. Provide the structure of propanoic anhydride.

    Answer: $(CH_3CH_2CO)_2O$ Section: 1

61. List the following carboxylic acid derivatives in decreasing order of reactivity in nucleophilic acyl substitution reactions: acetic anhydride, methyl acetate, acetamide, acetyl chloride.

    Answer: acetyl chloride > acetic anhydride > methyl acetate > acetamide

    Section: 5

62. Provide the proper IUPAC name for $(CH_3)_3CCH_2CH_2CH(CN)CH_3$.

    Answer: 2,5,5-trimethylhexanenitrile   Section: 1

MULTIPLE CHOICE

63. In which of the following compounds does the carbonyl stretch in the IR spectrum occur at the lowest wavenumber?

    a. cyclohexanone
    b. ethyl acetate
    c. propanoyl chloride
    d. pentanamide
    e. γ-butyrolactone

    Answer: d   Section: 2

64. Which of the following compounds has the lowest boiling point?

    a. 1-butanol
    b. butanoic acid
    c. butanenitrile
    d. methyl propanoate
    e. butanamide

    Answer: d   Section: 3

SHORT ANSWER

65. Provide the structure of the major organic product of the following reaction sequence.

    CO_2H

    1. $SOCl_2$, Δ
    2. $NaO_2CCH_2CH_3$

    Answer:

    Section: 18

66. In the lab, a student tried to prepare N,N-diethylacetamide by heating acetyl chloride with two equivalents of triethylamine. Was the student successful? Explain.

Answer:
No, the reaction was unsuccessful. Tertiary amines do not react with acid chlorides to form stable products because there is no H bonded to the nitrogen in the amine reagent which can be lost subsequent to amine attack on the carbonyl. The reaction would have proved successful had diethylamine been used.

Section: 7

67. When benzoyl chloride is reacted with ammonia to prepare benzamide, two equivalents of ammonia must be used. Explain why this is the case.

Answer:
The reaction of an acyl chloride with ammonia produces an amide and an H^+. The proton generated in the reaction will protonate unreacted ammonia and thereby render it nonnucleophilic.

Section: 7

MULTIPLE CHOICE

68. Which of the following is (are) formed in the reaction between ethyl butanoate and ethyl amine?

a. ethanol
b. 1-butanol
c. N-ethylbutanamide
d. both a and c
e. both b and c

Answer: d   Section: 9

69. Which of the following esters undergoes hydrolysis in base most easily?

a. $CH_3CH_2CO_2CH_3$
b. $(CH_3)_2CHCH_2CO_2CH_3$
c. $C_6H_5O_2CCH_3$
d. $p-CH_3C_6H_4O_2CCH_3$
e. $p-NO_2C_6H_4O_2CCH_3$

Answer: e   Section: 9

70. Predict the major organic product of the reaction between N-methylbenzamide and sodium propanoate.

    a. N-propylbenzamide
    b. propyl benzoate
    c. phenyl propanoate
    d. benzoic propanoic anhydride
    e. no reaction

    Answer: e Section: 13

Chapter 17:   Carbonyl Compounds II

MULTIPLE CHOICE

1. Which of the following compounds is an aldehyde?

   a. I
   b. II
   c. III
   d. IV
   e. V

Answer: d Section: 1

2. Which of the following compounds is a ketone?

   a. I
   b. II
   c. III
   d. IV
   e. V

I.

II.

III.

IV.

V.

Answer: b Section: 1

3. What is the common name for the following compound?

$$CH_2-\overset{\overset{\textstyle O}{\|}}{C}-H$$
$$|$$
$$Cl$$

   a. Chloroaldehyde
   b. α-Chloroacetaldehyde
   c. β-Chloroacetaldehyde
   d. 2-Chloroethanal
   e. α-Chloroethanal

Answer: b Section: 1

4. Which of the following compounds is acetone?

a.
$$CH_3-\overset{\overset{\displaystyle O}{\|}}{C}-CH_3$$

b.
$$CH_3-\overset{\overset{\displaystyle O}{\|}}{C}-H$$

c.
$$CH_3-O-\overset{\overset{\displaystyle O}{\|}}{C}-CH_3$$

d.
$$CH_3-\overset{\overset{\displaystyle O}{\|}}{C}-OH$$

e.
$$CH_3-\overset{\overset{\displaystyle O}{\|}}{C}-CH_2CH_3$$

Answer: a Section: 1

5. What is the IUPAC name for the following compound?

a. 2-Bromobutanal
b. α-Bromobutanal
c. 3-Bromobutanal
d. β-Bromobutyraldehyde
e. 3-Bromobutanone

Answer: c Section: 1

6. Which of the following compounds is benzaldehyde?

   a. I
   b. II
   c. III
   d. IV
   e. V

I.

II.

III.

IV.

V.

Answer: d   Section: 1

7. Which of the following is cyclohexanone?

    a. I
    b. II
    c. III
    d. IV
    e. V

    I.

    II.

    III.

    IV.

    V.

Answer: e Section: 1

8. What is the IUPAC name for the following compound?

$$CH_3\overset{\overset{\displaystyle O}{\|}}{C}\underset{\underset{\displaystyle CH_3}{|}}{C}HCH_2\underset{\underset{\displaystyle CH_3}{|}}{C}HCl$$

    a. 5-Chloro-3-methylhexanone
    b. 1-Chloro-1,3-dimethyl-4-pentanone
    c. 5-Chloro-3,5-dimethyl-2-pentanone
    d. 5-Chloro-3-methyl-2-hexanone
    e. 2-Chloro-4-methyl-5-hexanone

Answer: d Section: 1

9. Which of the following characterizes the reactions of aldehydes and ketones?

    a. Electrophilic addition
    b. Electrophilic substitution
    c. Nucleophilic acyl substitution
    d. Nucleophilic addition
    e. Free radical addition

Answer: d Section: 3

10. Which of the following reactions can be used to prepare 3-methyl-3-hexanol,

$$
\overset{\displaystyle OH}{\underset{\displaystyle CH_3}{CH_3CH_2\overset{|}{\underset{|}{C}}CH_2CH_2CH_3}} \quad ?
$$

  a.
$$CH_3CH_2\overset{\displaystyle O}{\overset{||}{C}}CH_2CH_2CH_3 \xrightarrow[\text{2. } H_3O^{\oplus}]{\text{1. } CH_3MgBr}$$

  b.
$$CH_3\overset{\displaystyle O}{\overset{||}{C}}CH_2CH_2CH_3 \xrightarrow[\text{2. } H_3O^{\oplus}]{\text{1. } CH_3CH_2MgBr}$$

  c.
$$CH_3\overset{\displaystyle O}{\overset{||}{C}}CH_2CH_3 \xrightarrow[\text{2. } H_3O^{\oplus}]{\text{1. } CH_3CH_2CH_2MgBr}$$

  d. a and b
  e. a, b, and c

Answer: e Section: 4

11. What is/are the <u>major</u> product(s) of the following reaction?

$$CH_3MgBr + HOCH_2\overset{\overset{\displaystyle O}{\|}}{C}H \longrightarrow \ ?$$

a.

$$HO-CH_2\overset{\overset{\displaystyle O}{\|}}{C}CH_3 + HMgBr$$

b.

$$HO-CH_2\overset{\overset{\displaystyle OMgBr}{|}}{\underset{\underset{\displaystyle CH_3}{|}}{C}}H$$

c.

$$CH_4 + BrMgOCH_2\overset{\overset{\displaystyle O}{\|}}{C}H$$

d.

$$BrMgCH_2\overset{\overset{\displaystyle O}{\|}}{C}H + CH_3OH$$

e.

$$HOCH_2\overset{\overset{\displaystyle O}{\|}}{C}CH_2MgBr + H_2$$

Answer: c Section: 4

12. Which of the following is the best method for preparing lactic acid from acetaldehyde?

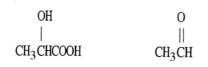

$\overset{\overset{\displaystyle OH}{|}}{CH_3}CHCOOH$	$CH_3\overset{\overset{\displaystyle O}{\|}}{C}H$
lactic acid	acetaldehyde

a. Cl_2; ^-OH; CH_3OH
b. CH_3MgBr; ^-OH
c. $KMnO_4$; ^-OH
d. $HC{\equiv}N$; H_3O^+/heat
e. Cl_2; Mg/ether; ^-OH

Answer: d Section: 4

13. Which of the following reagents can be used to reduce acetaldehyde to ethyl alcohol?

    a. 1. $LiAlH_4$/2. H_3O^+
    b. 1. $NaBH_4$/2. H_3O^+
    c. H_2/Pt
    d. a and d
    e. a, b, and c

    Answer: e Section: 5

14. Which of the following reactions can be used to prepare:

    a. I
    b. II
    c. III
    d. b and c
    e. a, b and c

    Answer: e Section: 4

15. Which of the following reactions is the best method for preparing acetaldehyde?

$$\begin{array}{c} O \\ \parallel \\ CH_3C\!-\!H \end{array}$$

a.
$$\begin{array}{c} O \\ \parallel \\ CH_3CCl \end{array} \xrightarrow[\text{2. } H_3O^{\oplus}]{\text{1. DIBAH}}$$

b.
$$\begin{array}{c} O \\ \parallel \\ CH_3CCl \end{array} \xrightarrow[\text{2. } H_3O^{\oplus}]{\text{1. LiAlH}_4}$$

c.
$$\begin{array}{c} O \\ \parallel \\ CH_3COH \end{array} \xrightarrow[\text{2. } H_3O^{\oplus}]{\text{1. LiAlH}_4}$$

d.
$$\begin{array}{c} O \\ \parallel \\ CH_3COCH_3 \end{array} \xrightarrow[\text{2. } H_3O^{\oplus}]{\text{1. NaBH}_4}$$

e.
$$\begin{array}{c} O \\ \parallel \\ CH_3CNH_2 \end{array} \xrightarrow[\text{2. } H_3O^{\oplus}]{\text{1. LiAlH}_4}$$

Answer: a   Section: 6

16. What is the <u>major</u> product of the following reaction?

   a. I
   b. II
   c. III
   d. IV
   e. V

I.

II.

III.

IV.

V.

Answer: b   Section: 6

17. Which of the following compounds is a hydrazone?

a. I
b. II
c. III
d. IV
e. V

I.   NHNH$_2$

II. N(CH$_3$)(CH$_3$)

III. —NH—NH—

IV. =N—NH$_2$

V.   NH$_2$

Answer: d   Section: 7

18. What is the major organic product of the following reaction?

$$CH_3-\overset{\overset{\displaystyle O}{\|}}{C}-H \ + \ HONH_2 \longrightarrow \ ?$$

a. CH$_3$CHOH
        |
       NHOH

b. CH$_3$C=O
        |
       NHOH

c. CH$_3$C=NOH
        |
        H

d. CH$_3$CHNHOCHCH$_3$
        | |
       OH   OH

e. CH$_3$CHOH
        |
       ONH$_2$

Answer: c   Section: 7

19. Which of the following is an enamine?

    a. I
    b. II
    c. III
    d. IV
    e. V

I.

$\quad$ N—CH$_3$

II.

$\quad$ NCH$_3$
$\quad$ CH$_3$

III.

$\quad$ CH$_3$
$\quad$ N
$\quad$ CH$_3$
$\quad$ OH

IV.

$\quad$ N=NH

V.

$\quad$ N—N

Answer: b Section: 7

20. What is the major organic product of the following reaction?

    a. I
    b. II
    c. III
    d. IV
    e. V

I.

II.

III.

IV.

V.

Answer: a   Section: 7

21. What is the major organic product of the following reaction?

    a. I
    b. II
    c. III
    d. IV
    e. V

I.

II.

III.

IV.

V.

Answer: e Section: 7

22. The compound:

$$\underset{\text{HOCH}_2\text{CH}_2\text{CH}_2\text{CH}_2\text{CH}}{\overset{\overset{\displaystyle O}{\|}}{}}$$

can form a hemiacetal by reacting with itself in solution. What is the structure of this hemiacetal?

a. I
b. II
c. III
d. IV
e. V

I.

II.

III.

IV. $\text{CH}_3\text{CH}_2\text{CH}_2\text{CH}_2\text{CH} \overset{\text{OH}}{\underset{\text{OH}}{}}$

V. $\text{CH}_3\text{CH}_2\text{CHOCH}_3$
              $\underset{\text{OH}}{|}$

Answer: b Section: 8

23. What sequence of reactions would best accomplish the following conversion?

$$\underset{\text{CH}_3\text{CCH}_2\text{CH}}{\overset{\overset{\displaystyle O \quad O}{\| \quad \|}}{}} \xrightarrow{\quad ? \quad} \underset{\text{CH}_3\text{CHCH}_2\text{CH}}{\overset{\overset{\displaystyle OH \quad O}{| \quad \|}}{}}$$

a. $\text{NaBH}_4 / \text{H}_3\text{O}^+$
b. $\text{HOCH}_2\text{CH}_2\text{OH} / \text{H}^+$; NaBH_4 ; $\text{H}^+ / \text{H}_2\text{O}$
c. $\text{KMnO}_4 / \text{HO}^-$; heat
d. $\text{LiAlH}_4 / \text{H}_3\text{O}^+$
e. H_2 / Pt ; PCC

Answer: b Section: 9

24. What is the <u>major</u> product of the following reaction?

    a. I
    b. II
    c. III
    d. IV
    e. V

I.

II.

III.

IV.

V.

Answer: c Section: 8

25. What carbonyl compound and what phosphonium ylide are required for the synthesis of the following alkene?

   a. I
   b. II
   c. III
   d. b or c
   e. a, b or c

I. $CH_3\overset{O}{\overset{\|}{C}}CH_3$ + (cyclohexylidene)$=P(C_6H_5)_3$

II. $CH_3CH_2\overset{O}{\overset{\|}{C}}H$ + (cyclohexylidene)$=P(C_6H_5)_3$

III. (cyclohexylidene)$=O$ + $CH_3CH_2CH=P(C_6H_5)_3$

Answer: d Section: 11

26. Which enantiomer is formed from attack of a methyl Grignard reagent on the si face of benzaldehyde:

    a. I
    b. II
    c. III
    d. IV
    e. V

I.

II.

III.

IV.

V.

Answer: a Section: 12

27. What reagent(s) would you use to accomplish the following conversion?

    a. CH_3Br; H_3O^+
    b. CH_3MgBr; H_3O^+
    c. $(CH_3)_2CuLi$; H_3O^+
    d. CH_3Br; $LiAlH_4$; H_3O^+
    e. $LiAlH_4$; CH_3MgBr; H_3O^+

Answer: b Section: 14

28. What reagent(s) would you use to accomplish the following conversion?

    a. CH_3Br; H_3O^+
    b. CH_3MgBr; H_3O^+
    c. $(CH_3)_2CuLi$; H_3O^+
    d. CH_3Br; $LiAlH_4$; H_3O^+
    e. $LiAlH_4$; CH_3MgBr; H_3O^+

Answer: c   Section: 14

SHORT ANSWER

29. Propose an IR frequency that would distinguish between the following compounds:

Answer:

Compound I shows a C-H peak for the aldehyde at ~2700 cm^{-1}, while compound II shows a peak at ~1380 cm^{-1} for CH_3.

Section: 99

30. Identify the compound whose molecular formula and proton NMR spectra are given below:

    $C_{15}H_{14}O$; a) triplet, at 2.1 ppm (2H)
    b) triplet, at 2.4 ppm (2H)
    c) multiplet, at 7.9 – 7.2 ppm (10H)

Answer:

Section: 99

31. An unknown compound (A), $C_{10}H_{12}O$, gave the following proton NMR data:
    (a) triplet, at 1.0 ppm (3H)
    (b) quartet, at 2.4 ppm (2H)
    (c) singlet, at 3.7 ppm (2H)
    (d) multiplet, at 7.2 ppm (5H)
        Propose a structure for (A)

    Answer:

32. Complete the following reaction sequence by supplying the missing information:

$$CH_3OH \xrightarrow{?} CH_3Br \xrightarrow{?} ? \xrightarrow[2.?]{1.?} CH_3CH_2OH$$

    Answer:

Section: 4

33. Give the products obtained from the following Wittig reaction. Show both stereoisomers.

    Answer:

Section: 11

34. Give three different sets of reagents (a carbonyl compound and a Grignard reagent) that could be used to prepare the following compound.

$$CH_3-\underset{\underset{CH_3}{|}}{\overset{\overset{OH}{|}}{C}}-CH_2CH_3$$

Answer:

a.

$$CH_3CH_2\overset{\overset{O}{\|}}{C}OCH_3 \text{ and } 2 \ CH_3MgBr$$

b.

$$CH_3\overset{\overset{O}{\|}}{C}CH_2CH_3 \text{ and } CH_3MgBr$$

c.

$$CH_3\overset{\overset{O}{\|}}{C}CH_3 \text{ and } CH_3CH_2MgBr$$

Section: 4

35. Propose a plausible mechanism that would explain the following conversions:

a)

b)

Answer:

a)

b)

Section: 7

36. Would you expect the carbonyl carbon of benzaldehyde to be more or less electrophilic than that of acetaldehyde? Explain using resonance structures.

    Answer:
    The carbonyl carbon of benzldehyde would be less electrophilic. Delocalization reduces its partial positive charge as seen in the resonance forms below.

    Section: 2

37. Provide the proper IUPAC name for $PhCH_2CH(CH_3)CH_2CH_2CHO$.

    Answer: 4-methyl-5-phenylpentanal   Section: 1

38. Provide the proper IUPAC name for $CH_3CHOHCH_2COCH_2C(CH_3)_2CH_2CH_3$.

    Answer: 2-hydroxy-6,6-dimethyl-4-octanone   Section: 1

39. Provide the structure of cyclohexanone oxime.

    Answer:

    Section: 7

40. Provide the structure of the diethyl acetal of butanal.

    Answer: $CH_3CH_2CH_2CH(OCH_2CH_3)_2$ Section: 8

MULTIPLE CHOICE

41. In carbon NMR, the carbon atom of the carbonyl group in aldehydes and ketones has a chemical shift of about:

   a. 20 ppm.
   b. 40 ppm.
   c. 60 ppm.
   d. 120 ppm.
   e. 200 ppm.

   Answer: e   Section: 99

42. In the proton NMR spectra of aldehydes and ketones, the protons bonded to carbons adjacent to the carbonyl group typically fall into which of the chemical shift ranges below?

   a. 1.0-2.0 ppm
   b. 2.0-3.0 ppm
   c. 4.0-4.5 ppm
   d. 7.0-8.0 ppm
   e. 9.0-10.0 ppm

   Answer: b   Section: 99

SHORT ANSWER

43. By which single-step process can benzene be readily converted into acetophenone?

   Answer: CH_3COCl, $AlCl_3$ - a Friedel-Crafts acylation reaction   Section: 99

44. What reagent can be used to convert benzophenone into triphenylmethanol?

   Answer: PhMgBr followed by hydrolysis   Section: 4

45. Beginning with sodium acetylide (NaCCH), propose a three-step synthesis of hexanal.

   Answer:
   1. $CH_3CH_2CH_2CH_2Br$
   2. Sia_2BH
   3. H_2O_2, NaOH
   Section: 4

46. Provide the major organic product which results when $(CH_3)_2CHCH_2CH_2COCl$ is treated with DIBAH.

   Answer: $(CH_3)_2CHCH_2CH_2CHO$ Section: 6

47. Provide a detailed, stepwise mechanism for the acid-catalyzed reaction of 2-butanone with ethylene glycol ($HOCH_2CH_2OH$) to produce an acetal.

Answer:

Section: 8

48. Provide a detailed, stepwise mechanism for the acid-catalyzed condensation reaction between benzaldehyde and methylamine.

Answer:

Section: 7

49. Provide a detailed, stepwise mechanism for the acid-catalyzed condensation reaction between cyclohexanone and H₂NOH.

Answer:

Section: 7

MULTIPLE CHOICE

50. When the carbonyl group of a neutral ketone is protonated,

    a. the resulting species becomes more electrophilic.
    b. the resulting species is activated toward nucleophilic attack.
    c. subsequent nucleophilic attack on the resulting species is said to occur under acid-catalyzed conditions.
    d. the resulting species has a positive charge.
    e. all of the above.

Answer: e   Section: 7

SHORT ANSWER

51. Draw the two major resonance forms of the cation which results when cyclohexanone's carbonyl group is protonated.

Answer:

Section: 7

52. Propose a synthesis of 4-phenyl-2-butanol from 3-phenylpropanal.

Answer:
1. CH_3MgBr
2. H_3O^+
Section: 4

53. Propose a synthesis of 3-methyl-4-heptyn-3-ol from 1-butyne.

Answer:
1. $NaNH_2$
2. $CH_3COCH_2CH_3$
3. H_3O^+
Section: 4

54. Provide the structure of the hydrate of cyclopentanone.

Answer:

Section: 8

55. Conversion of aldehydes and ketones to imines is an acid-catalyzed process. Explain why this conversion is actually hindered by the presence of too much acid.

Answer:
At very low pH, much more of the amine is protonated and can no longer function as a nuclepohile.
Section: 7

56. Provide the major organic product of the reaction of aniline with 3-pentanone.

Answer:

Section: 7

57. Propose a sequence of steps to carry out the following conversion.

Answer:
1. $HOCH_2CH_2OH$, H^+
2. $NaNH_2$
3. CH_3I
4. H_3O^+
5. $NaBH_4$ <u>or</u> $LiAlH_4$

Section: 9

58. When $HOCH_2CH_2CH_2CH_2COCH_2CH_2CH_2CH_2OH$ is heated in the presence of an acid catalyst, a reaction occurs. The product has the formula $C_9H_{16}O_2$. Provide the structure of this product.

Answer:

Section: 8

59. Propose a sequence of steps to carry out the following conversion.

Answer:
1. $HOCH_2CH_2OH$, H^+
2. $CH_3CH_2CH_2MgBr$
3. H_3O^+

Section: 9

409

60. Which would be more appropriate as the reduction in the following sequence, a Clemmensen or a Wolff-Kishner? Explain.

Answer:
Wolff-Kishner. The acidic conditions of the Clemmensen would remove the protecting group causing reduction of the aldehyde as well.
Section: 7

61. Which carbonyl is more susceptible to nucleophilic attack, that of cyclohexanone or that of hexanal? Provide two reasons for your choice.

Answer:
Hexanal. The partial positive charge on the carbonyl carbon causes carbonyl compounds to be attacked by nucleophiles. Because a hydrogen is electron-withdrawing relative to an alkyl group, an aldehyde has a greater partial positive charge on its carbonyl carbon and is thus more susceptible to nucleophilic attack. Steric factors also contribute to the greater reactivity of an aldehyde. Because the hydrogen attached to the carbonyl carbon of an aldehyde is smaller than the alkyl group attached to the carbonyl carbon of a ketone, the carbonyl carbon of an aldehyde is more accessible to nucloephilic attack.
Section: 2

62. List the following carbonyl compounds in order of decreasing reactivity toward nucleophiles: ester, acid chloride, amide, aldehyde, ketone.

Answer: acid chloride > aldehyde > ketone > ester > amide   Section: 2

MULTIPLE CHOICE

63. Why do aldehydes undergo nucleophilic addition reactions while esters undergo nucleophilic acyl substitution reactions?

   a. The carbonyl carbon of an ester is more electrophilic than that of an aldehyde.
   b. Aldehydes are more sterically hindered than esters.
   c. Once the nucleophile adds to an aldehyde, the tetrahedral intermediate is too sterically hindered to eliminate one of the attached groups.
   d. The ester carbonyl carbon is sp3 hybridized while the aldehyde carbonyl carbon is sp2 hybridized.
   e. Once the nucleophile adds to an aldehyde, neither H- nor R- can be eliminated since they are strongly basic.

   Answer: e   Section: 3

64. What is the major product when $(CH_3)_2CHCH_2CH_2Br$ is subjected to the following reaction sequence: 1. Mg, ether 2. CO_2 3. H_3O^+?

   a. 3-methyl-1-butanol
   b. 2-methyl-2-butanol
   c. 4-methyl-1-pentanol
   d. 4-methylpentanoic acid
   e. 3-methylbutanal

   Answer: d   Section: 4

SHORT ANSWER

65. Propose a sequence of steps to convert propyne to 4-heptanol.

   Answer:
   1. $NaNH_2$
   2. $CH_3CH_2CH_2CHO$
   3. H_2 (excess), Pt
   Section: 4

MULTIPLE CHOICE

66. Which of the sequences works best to accomplish the following conversion?

    a. 1. NaCN, HCl   2. H_2, Pt
    b. 1. H_2NCH_2MgBr 2. H_3O^+
    c. 1. $NaNH_2$ 2. H_3O^+
    d. 1. H_2NNH_2, H^+ 2. H_3O^+
    e. 1. NH_3, H^+ 2. H_2, Pt

    Answer: a   Section: 4

67. Which of the following amines will react with cyclopentanone to form an enamine?

    a. $CH_3CH_2CH_2CH_2NH_2$
    b. $(CH_3)_3N$
    c. pyridine
    d. $(CH_3)_3CNH_2$
    e. none of the above

    Answer: e   Section: 7

68. Which of the following synthetic routes works best for preparing 4-octene?

    a. $CH_3CH_2CH_2CH=PPh_3$ + $CH_3CH_2CH_2CHO$
    b. 4-bromooctane + $NaOCH_3$
    c. 4-bromooctane + $KOC(CH_3)_3$
    d. $CH_3CH_2CH_2CH_2MgBr$ + $CH_3CH_2CH_2CHO$
    e. $CH_3CH_2CH=PPh_3$ + $CH_3CH_2CH_2CH_2CHO$

    Answer: a   Section: 11

SHORT ANSWER

69. Provide the sequence of reagents by which a phosphonium ylide can be prepared from 1-bromobutane.

    Answer:
    1. Ph_3P
    2. butyllithium

    Section: 11

70. When camphor undergoes nucleophilic attack by CH_3CH_2MgBr, the major product is the one in which the Grignard reagent has added endo. Explain the stereoselectivity observed. Which face of the carbonyl carbon, re or si, was attacked?

1. CH_3CH_2MgBr

2. H_3O^+

Answer:
The Grignard reagent attacked the more sterically accessible si face of the carbonyl carbon.

Section: 12

Chapter 18: Oxidation-Reduction Reactions

MULTIPLE CHOICE

1. Which of the following is <u>not</u> a reduction reaction?

   a. I
   b. II
   c. III
   d. IV
   e. V

I. $R-\overset{\overset{\textstyle O}{\|}}{C}-Cl \xrightarrow[\text{Lindlar's Catalyst}]{H_2} CH_3-\overset{\overset{\textstyle O}{\|}}{C}-H$

II. $CH_2{=}CH_2 \xrightarrow[CCl_4]{Br_2} \underset{\underset{\textstyle Br \quad Br}{|\quad|}}{CH_2-CH_2}$

III. $CH_3-\overset{\overset{\textstyle O}{\|}}{C}-CH_3 \xrightarrow[\text{OH, heat}]{NH_2NH_2} CH_3CH_2CH_3$

IV. $R-\overset{\overset{\textstyle O}{\|}}{C}-R \xrightarrow[\text{2. } H_3O^+]{\text{1.} NaBH_4} R-\underset{\underset{\textstyle OH}{|}}{CH}-R$

V. $\xrightarrow[HCl]{Zn(Hg)}$

Answer: b Section: 1

414

2. What is the major organic product of the following reaction?

    a. I
    b. II
    c. III
    d. IV
    e. V

Answer: c Section: 1

3. Which of the following reducing agents can cause the formation of cis-2-butene as a major product of the reaction shown below?

    a. H_2/Lindlar's Catalyst
    b. Na/NH_3
    c. Li/NH_3
    d. 1. $LiAlH_4$ /2. H_3O^+
    e. 1. $NaBH_4$ /2. H_3O^+

Answer: a Section: 1

4. Which of the following carbonyl compounds can be reduced by H_2/Raney Ni?

   a.   O
   $$CH_3-\overset{\displaystyle \parallel}{C}-OH$$
   b.   O
   $$CH_3-\overset{\displaystyle \parallel}{C}-OCH_3$$
   c.   O
   $$CH_3-\overset{\displaystyle \parallel}{C}-NHCH_3$$
   d.   O
   $$CH_3-\overset{\displaystyle \parallel}{C}-Cl$$
   e.   O
   $$CH_3-\overset{\displaystyle \parallel}{C}-NH_2$$

   Answer: d   Section: 1

5. Which of the following reagents gives the reaction shown?

   a. H_2/H_2SO_4
   b. H_2O/Ni
   c. H_2O/H_2SO_4
   d. H^+/H_2O
   e. H_2/Ni

   Answer: e   Section: 1

6. Which of the following reagents would reduce carboxylic acids and esters into alcohols?

   a. H_2/Raney Ni
   b. $1.LiAlH_4/2.H_3O^+$
   c. Na/NH_3
   d. $1.NaBH_4/2.H_3O^+$
   e. $Zn(Hg)/H^+$

   Answer: b   Section: 1

7. Which of the following reducing agents is best used in the reaction shown below?

$$CH_3-\overset{\overset{\textstyle O}{\|}}{C}-NH_2 \xrightarrow{\quad ? \quad} CH_3CH_2NH_2$$

a. H_2/Raney Ni
b. 1.$LiAlH_4$/2.H_3O^+
c. Na/NH_3
d. 1.$NaBH_4$/2.H_3O^+
e. Zn(Hg)/H^+

Answer: b   Section: 1

8. Which of the following reducing agents is best used in the reaction shown below?

$$CH_2=CH-\overset{\overset{\textstyle O}{\|}}{C}-CH_3 \xrightarrow{\quad ? \quad} CH_2=CH-\overset{\overset{\textstyle OH}{|}}{C}H-CH_3$$

a. H_2/Raney Ni
b. 1.$LiAlH_4$/2.H_3O^+
c. Na/NH_3
d. 1.$NaBH_4$/2.H_3O^+
e. Zn(Hg)/H^+

Answer: d   Section: 1

9. Which of the following alcohols is oxidized to a ketone by chromic acid?

   a. I
   b. II
   c. III
   d. IV
   e. V

   I.   OH

   II. OH

   III. CH$_2$OH

   IV. CH$_2$OH

   V.

         CH$_3$
         |
         C—CH$_3$
         |
         OH

   Answer: b Section: 2

10. Which of the following reagents does <u>not</u> form an alcohol from an aldehyde as shown below?

$$R-\overset{\overset{\textstyle O}{\|}}{C}-H \xrightarrow{\ ?\ } R-CH_2-OH$$

   a. H$_2$/Pt
   b. 1.NaBH$_4$/2.H$_3$O$^+$
   c. 1.LiAlH$_4$/2.H$_3$O$^+$
   d. 1.Ag$_2$O,NH$_3$/2.H$_3$O$^+$
   e. H$_2$/Raney Ni

   Answer: d Section: 1

11. Which of the following reagents <u>does</u> form an alcohol from a ketone as shown below?

    a. $NH_2NH_2/^-OH$
    b. $Zn(Hg)/H^+$
    c. $Ag(NH_3)_2NO_3$
    d. $K_2Cr_2O_7/H^+$
    e. $H_2/Raney\ Ni$

Answer: e Section: 1

12. Which of the following oxidizing agents is used in a Baeyer-Villiger oxidation?

    a. $NH_2NH_2/^-OH$
    b. $Zn(Hg)/H^+$
    c. $Ag(NH_3)_2NO_3$
    d. $K_2Cr_2O_7/H^+$
    e. $\overset{O}{\underset{\|}{RCOOH}}$

Answer: e Section: 3

13. What is the major organic product of the following reaction?

   a. I
   b. II
   c. III
   d. IV
   e. V

Answer: b Section: 3

14. What is the <u>major</u> organic product of the following reaction?

   a. I
   b. II
   c. III
   d. IV
   e. all of the above

Answer: a Section: 5

15. Which of the following reagents effectively cleaves a carbon-carbon double bond?

   a. $Cl_2/h\upsilon$
   b. $R-\overset{\overset{\displaystyle O}{||}}{C}-OOH$
   c. H_2SO_4
   d. cold $KMnO_4$
   e. hot $KMnO_4$

Answer: e Section: 7

16. An unknown alkene was treated with ozone, and then with H_2O_2 forming the following product:

a. I
b. II
c. III
d. IV
e. V

$$CH_3-\overset{\overset{\displaystyle O}{\|}}{C}-CH_2-CH_2-CH_2-CH_2-\overset{\overset{\displaystyle O}{\|}}{C}-OH$$

What is the structure of the unknown?

I.

II.

III.

IV.

V.

Answer: d   Section: 7

SHORT ANSWER

17. Is it easier to reduce a carbon-oxygen double bond or a carbon-carbon double bond using catalytic hydrogenation?

Answer: A carbon-carbon double bond is easier to reduce.   Section: 1

MULTIPLE CHOICE

18. What is the major organic product(s) when the following molecule is treated with ozone, and then with Zn/H_2O?

    a. I
    b. II
    c. III
    d. IV
    e. V

I.

, $H-\overset{\overset{O}{\|}}{C}-H$

II.

, CO_2, H_2O

III.

, CH_2O

IV.

, $H-\overset{\overset{O}{\|}}{C}-OH$

V.

, $HO-\overset{\overset{O}{\|}}{C}-OH$

Answer: a   Section: 7

423

19. Which of the following reagents best facilitates the reaction shown below?

    a. I
    b. II
    c. III
    d. IV
    e. V

    I. $1.O_3/ 2.Zn, H_2O$

    II. $1.O_3/ 2.H_2O_2$

    III. $CH_3-C-OOH$
            $\quad\quad\quad \underset{O}{\|}$

    IV. $KMnO_4$ cold

    V. $KMnO_4$ hot

    Answer: c   Section: 4

20. What is the major product when the following compound reacts with MMPP?

    a. I
    b. II
    c. III
    d. IV
    e. V

Answer: e Section: 4

21. Which of the following reagents is <u>not</u> an oxidizing agent?

    a. $KMnO_4$
    b. CrO_3
    c. NADH
    d. PCC
    e. HIO_4

Answer: c Section: 10

22. Which of the following reagents can be used to oxidize 1° alcohols to aldehydes?

    a. $KMnO_4$
    b. MnO_2
    c. $K_2Cr_2O_7$
    d. H_2O_2
    e. PCC

Answer: e Section: 2

23. Which of the following reagents is best used for the conversion shown below?

    a. 1.NaBH$_4$/2.D$_3$O$^+$
    b. 1.NaBD$_4$/2.H$_3$O$^+$
    c. 1.LiAlH$_4$/2.D$_3$O$^+$
    d. 1.H$_2$/Pt/2.D$_2$O
    e. 1.LiAlH$_4$/2.H$_3$O$^+$

Answer: b   Section: 1

SHORT ANSWER

24. What is the product of the following reduction reaction?

Answer:

The product is

Section: 1

25. When pyruvic acid is reduced using 1.NaBH$_4$/2.H$_3$O$^+$, what is the expected product?

Answer:
NaBH$_4$ can reduce only the keto group. The product is lactic acid.

Section: 1

26. Would you expect alkenes and alkynes to be reduced by $NaBH_4$? Why?

Answer:
No, alkenes and alkynes cannot be reduced by $NaBH_4$ because they do not possess a partial positive charge.
Section: 1

27. What is PCC and what is it used for?

Answer:
PCC is pyridinium chlorochromate, a chromic anhydride (CrO_3) and an equivalent of HCl dissolved in pyridine. It is a mild oxidizing agent used to convert alcohols to aldehydes.
Section: 2

28. Propose a mechanism for the following oxidation reaction known as the Baeyer-Villiger Oxidation:

Answer:

Section: 3

29. Give the structure of the alkyne that gives the following products upon ozonolysis.

$$HO-\overset{\overset{\displaystyle O}{\|}}{C}-CH_2CH_2-\overset{\overset{\displaystyle O}{\|}}{C}-OH, \quad 2 \ CH_3-\overset{\overset{\displaystyle O}{\|}}{C}-OH$$

Answer: $CH_3C{\equiv}C-CH_2CH_2-C{\equiv}CCH_3$ Section: 8

30. What does MMPP stand for? Give the structure of MMPP.

Answer:
It stands for magnesium monoperoxyphthalate.

It is a stable molecule that is used for epoxidation.

Section: 4

MULTIPLE CHOICE

31. Classify the reaction below as an oxidation, a reduction, or neither.

$$(CH_3)_2CHCH_2OH \longrightarrow (CH_3)_2CHCHO$$

a. oxidation
b. reduction
c. neither

Answer: a   Section: 98

32. Classify the reaction below as an oxidation, a reduction, or neither.

$$PhCO_2H \longrightarrow PhCH_2OH$$

a. oxidation
b. reduction
c. neither

Answer: b   Section: 98

33. Classify the reaction below as an oxidation, a reduction, or neither.

cis-2-pentene $\longrightarrow$ pentane

a. oxidation
b. reduction
c. neither

Answer: b Section: 98

SHORT ANSWER

34. Provide the structure of the major organic product in the reaction below.

Answer:

Section: 2

35. Provide the structure of the major organic product in the reaction below.

Answer:

Section: 2

36. Provide the structure of the major organic product in the reaction below.

$$\text{PCC}$$

Answer: No reaction. Tertiary alcohols are not oxidized by PCC.

Section: 2

37. Provide the structure of the major organic product in the reaction below.

$$(CH_3)_3CCH_2CH_2OH \xrightarrow[H_2SO_4]{Na_2Cr_2O_7}$$

Answer: $(CH_3)_3CCH_2CO_2H$ Section: 2

38. Provide the structure of the major organic product in the reaction below.

$$\xrightarrow{Na_2Cr_2O_7 H_2SO_4}$$

Answer:

Section: 2

MULTIPLE CHOICE

39. In the chromic acid oxidation of alcohols, the chromium is:

a. reduced from Cr^{+6} to Cr^{+3}.
b. oxidized from Cr^{+6} to Cr^{+3}.
c. reduced from Cr^{+3} to Cr^{+6}.
d. oxidized from Cr^{+3} to Cr^{+6}.
e. none of the above.

Answer: a Section: 2

SHORT ANSWER

40. Chromic acid oxidations are believed to proceed through a chromate ester intermediate. In the oxidation of 1-propanol, provide the structure of this ester and give the mechanism for its conversion to the intermediate aldehyde.

    Answer:

    Section: 2

41. The enzyme alcohol dehydrogenase, which is produced by the liver, converts ethanol to _____, a normal body metabolite.

    Answer: acetic acid Section: 10

42. Poisonings by methanol and ethylene glycol are relatively common. How are these poisonings treated clinically?

    Answer:
    The enzyme alcohol dehydrogenase converts these compounds to toxic substances in the body. If the patient receives intravenous infusions of ethanol, then the ethanol swamps the enzyme and the potentially deadly alcohols pass from the body unoxidized.
    Section: 10

43. Provide the structure of the major organic product in the reaction below.

    $$CH_3CH_2CH(CH_3)CH_2CO_2CH_3 \xrightarrow[\text{2. } H_3O^+]{\text{1. } LiAlH_4}$$

    Answer: $CH_3CH_2CH(CH_3)CH_2CH_2OH$ Section: 1

44. What series of synthetic steps could be used to prepare 1-methylcyclohexanol from cyclohexene?

    Answer:
    1. H_2O, H^+ or BH_3; H_2O_2, ^-OH or $Hg(OAc)_2$, H_2O; $NaBH_4$
    2. $NaCr_2O_7$, H^+ or PCC or $KMnO_4$
    3. CH_3MgI
    4. H_3O^+
    Section: 2

45. What series of synthetic steps could be used to prepare 3-hexanol from 1-propanol?

    Answer:
    1. PCC
    2. $CH_3CH_2CH_2MgBr$
    3. H_3O^+
    or
    1. HBr or PBr_3
    2. Mg, Et_2O
    3. CH_3CH_2CHO
    Section: 2

46. What series of synthetic steps could be used to prepare $(CH_3)_2CHCHO$ from isobutane?

    Answer:
    1. Br_2, hυ
    2. NaOEt, heat
    3. BH_3
    4. H_2O_2, ^-OH
    5. PCC
    Section: 2

47. What series of synthetic steps could be used to prepare $PhCHOHCH_2CH_3$ from $PhCH_2OH$?

    Answer:
    1. PCC
    2. CH_3CH_2MgBr
    3. H_3O^+
    Section: 2

48. What series of synthetic steps could be used to carry out the transformation shown below?

    Answer:
    1. BH_3
    2. H_2O_2, ^-OH
    3. PCC
    Section: 2

49. Draw the major organic product generated in the reaction below. Pay particular attention to regio- and stereochemical detail.

$$\text{H}_2, \text{Pt}$$

Answer:

Section: 1

50. Draw the major organic product generated in the reaction below. Pay particular attention to regio- and stereochemical detail.

CH₃

1. CH₃CO₃H
2. H⁺, H₂O

Answer:

OH

''''''|| CH₃   + enantiomer

'''''|| OH

Section: 4

51. Draw the major organic product generated in the reaction below. Pay particular attention to regio- and stereochemical detail.

PhCO₃H

CH₂Cl₂

Answer:

O

H''''''   '''''|CH₂CH₃   + enantiomer

CH₃   H

Section: 4

433

52. Draw the major organic product generated in the reaction below. Pay particular attention to regio- and stereochemical detail.

Answer:

+ enantiomer

Section: 5

53. Draw the major organic product generated in the reaction below. Pay particular attention to regio- and stereochemical detail.

Answer:

Section: 5

54. Draw the major organic product generated in the reaction below. Pay particular attention to regio- and stereochemical detail.

Answer:

Section: 7

55. Draw the major organic product generated in the reaction below. Pay particular attention to regio- and stereochemical detail.

$$\xrightarrow[\text{(warm, conc.)}]{\text{KMnO}_4}$$

Answer:

Section: 7

56. Draw the major organic product generated in the reaction below. Pay particular attention to regio- and stereochemical detail.

$$\xrightarrow[\text{2. (CH}_3)_2\text{S}]{\text{1. O}_3}$$

Answer:

+ CH_3CHO

Section: 7

57. Provide the reagents necessary to complete the following transformation.

Answer:
1. NaOCH$_3$, CH$_3$OH
2. O$_s$O$_4$, H$_2$O$_2$ or cold, dilute KMnO$_4$, $^-$OH
Section: 5

58. Give the structure of the alkene which would yield the following products upon ozonolysis-reduction.

$CH_3CH_2CH_2CH_2CHO$ + CH_2O

Answer: $CH_3CH_2CH_2CH_2CH_2=CH_2$ Section: 7

435

59. Give the structure of the alkene which would yield the following products upon ozonolysis-reduction.

CH_3COCH_3 + CH_3CH_2CHO

Answer: $(CH_3)_2C=CHCH_2CH_3$ Section: 7

60. β-Ocimene is a natural product with a pleasant odor. Based on the information below, deduce the structure of β-ocimene.

$$\beta\text{-Ocimene} \xrightarrow{H_2, PT} 2,6\text{-dimethyloctane}$$
$$(C_{10}H_{16})$$

$$\xrightarrow[\text{2. } (CH_3)_2S]{\text{1. } O_3} CH_2O + CH_3COCH_3 + CH_3COCHO$$
$$+ OHCCH_2CHO$$

Answer:

Section: 7

61. Provide the structure of the major organic product of the following reaction sequence.

1. NaCCCH$_3$

2. H$_3$O$^+$

3. NA, NH$_3$

Answer:

Section: 1

62. Provide the structure of the major organic product of the following reaction sequence.

$$\xrightarrow[\text{2. H}_2\text{O}]{\text{1. NaBH}_4}$$

Answer:

Section: 1

63. Provide the structure of the major organic products of the following reaction sequence.

$$\xrightarrow[\text{2. H}_2\text{O}]{\text{1. LiAlH}_4}$$

Answer:

(racemic)   + $CH_3CH_2CH_2CH_2OH$

Section: 1

437

MULTIPLE CHOICE

64. When the optically active alcohol below is oxidized by H_2CrO_4, the major product is _____.

    a. a ketone which is meso and optically inactive
    b. an optically active ketone
    c. a racemic mixture of ketones which is optically inactive
    d. an aldehyde which is meso and optically active
    e. an acyclic carboxylic acid

    Answer: b   Section: 2

65. What compound results when (R)-2-hexanol undergoes a Swern oxidation?

    a. 2-hexanone
    b. 3-hexanone
    c. hexanoic acid
    d. pentanoic acid
    e. hexanal

    Answer: a   Section: 2

SHORT ANSWER

66. Propose a synthesis of 3-hexanone using only 1-bromopropane as the source of carbon.

    Answer:

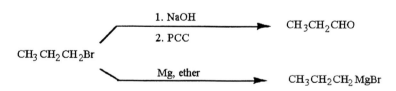

    Section: 2

MULTIPLE CHOICE

67. What alkene would you treat with RCO₃H in order to obtain the compound below and its enantiomer?

    a. (Z)-2-methyl-2-pentene
    b. (E)-2-methyl-2-pentene
    c. (Z)-3-methyl-2-pentene
    d. (E)-3-methyl-2-pentene
    e. 3-methyl-1-pentene

Answer: c   Section: 4

68. Treatment of 1,2-diethylcyclopentene with OsO₄ produces _____.

    a. a racemic mixture of osmates
    b. a meso osmate
    c. a racemic mixture of trans vicinal diols
    d. a meso trans vicinal diol
    e. an optically active trans vicinal diol

Answer: b   Section: 5

SHORT ANSWER

69. An alkene is treated with OsO₄ followed by aqueous NaHSO₃ and the resulting product is treated with periodic acid. The product mixture of this sequence contains an equimolar mixture of cyclohexanone and cyclopentanone. What is the structure of the alkene? What compounds would be produced if this alkene were subjected to ozonolysis followed by a zinc work-up?

Answer:
When this alkene is ozonolyzed, a 1:1 molar mixture of cyclohexanone and cyclopentanone is generated.

Section: 7

MULTIPLE CHOICE

70. What organic compound is produced when 3-hexyne undergoes ozonolysis followed by hydrolysis?

    a. 3,4-hexanedione
    b. <u>cis</u>-3,4-hexanediol
    c. <u>trans</u>-3,4-hexanediol
    d. propanal
    e. propanoic acid

    Answer: e Section: 8

Chapter 19:   Carbonyl Compounds III

MULTIPLE CHOICE

1. Which of the following compounds does <u>not</u> have an alpha hydrogen?

   a. I
   b. II
   c. III
   d. IV
   e. all of the above

   I.
   $$\text{C}_6\text{H}_5-\overset{\overset{\displaystyle O}{\|}}{C}H$$

   II.
   $$H-\overset{\overset{\displaystyle O}{\|}}{C}-H$$

   III.

   IV.
   $$\text{CH}_3\overset{\overset{\displaystyle CH_3}{|}}{\underset{\underset{\displaystyle CH_3}{|}}{C}}-\overset{\overset{\displaystyle O}{\|}}{C}H$$

   Answer: e Section: 1

2. Which of the following underlined alpha hydrogens are the <u>most</u> acidic?

   a.
   $$\underline{\text{CH}_3}-\overset{\overset{\displaystyle O}{\|}}{C}H$$

   b.
   $$\text{CH}_3-\overset{\overset{\displaystyle O}{\|}}{C}-\underline{\text{CH}_2}-\overset{\overset{\displaystyle O}{\|}}{C}H$$

   c.
   $$\text{CH}_3-\overset{\overset{\displaystyle O}{\|}}{C}-\underline{\text{CH}_2}-\overset{\overset{\displaystyle O}{\|}}{C}-\text{OCH}_3$$

   d.
   $$\underline{\text{CH}_3}-\overset{\overset{\displaystyle O}{\|}}{C}-\text{CH}_3$$

   e.
   $$\underline{\text{CH}_3}-\overset{\overset{\displaystyle O}{\|}}{C}-\text{O}-\text{CH}_3$$

   Answer: b Section: 1

441

3. Which of the labeled hydrogen atoms in the following structure is the <u>most</u> acidic?

    a. 1
    b. 2
    c. 3
    d. 4
    e. 5

Answer: e Section: 1

4. Which of the labeled hydrogen atoms in the following structure is the <u>most</u> acidic?

    a. 1
    b. 2
    c. 3
    d. 4
    e. 5

Answer: c Section: 1

5. Which of the labeled hydrogen atoms in the following structure is the <u>most</u> acidic?

    a. 1
    b. 2
    c. 3
    d. 4
    e. 5

Answer: d Section: 1

6. Which of the following is a β-diester?

a.

$$CH_3OCCH_2COCH_3$$

b.

$$CH_3CCH_2CCH_3$$

c.

$$CH_3CCH_2COCH_3$$

d.

$$CH_3OCOCH_3$$

e.

$$CH_3OCOCH$$

Answer: a Section: 1

7. When compared to the keto form, the enol form of which of the following compounds is <u>most</u> stable?

    a. I
    b. II
    c. III
    d. IV
    e. V

Answer: c Section: 2

8. Which of the following is the most stable enol form of 1,3-cyclohexanedione?

   a. I
   b. II
   c. III
   d. IV
   e. V

Answer: e

I.

II.

III.

IV.

V.

Section: 2

9. How would you accomplish the following conversion?

   a. H^+/H_2O; $HC≡N$
   b. $-OH/NaC≡N$
   c. Br_2/H^+, H_2O; $NaC≡N$
   d. HBr; $NaC≡N$
   e. $LiAlH_4$; $HC≡N$; $KMnO_4$

Answer: c Section: 4

445

10. Which of the following compounds will give a positive iodoform test?

    a. propanal
    b. 2-pentanone
    c. 3-pentanone
    d. benzophenone
    e. cyclohexanone

Answer: b   Section: 4

11. Which is the <u>major</u> organic product from the following reaction?

$$CH_3CH_2COH \xrightarrow[\text{2. } H_3O^+]{\text{1. } Br_2, PBr_3}$$ (with C=O, shown as O with || above COH)

a.
$$BrCH_2CH_2COH$$ (with O, || above the C of COH)

b.
$$CH_3CH_2CBr$$ (with O, || above the C of CBr)

c.
$$CH_3CHCOH$$ (with O, || above the C of COH; Br below the CH)

d.
$$CH_3CH_2COBr$$ (with O, || above the C of COBr)

e.
$$CH_3CH_2COH$$ (with OH above the final C; Br below it)

Answer: c   Section: 5

12. What is the <u>major</u> organic product of the following reaction?

    a. I
    b. II
    c. III
    d. IV
    e. V

I.

II.

III.

IV.

V.

Answer: d Section: 8

13. What is the <u>major</u> organic product of the following reaction?

    a. I
    b. II
    c. III
    d. IV
    e. V

Answer: a   Section: 10

14. Which of the following methods can be used to prepare the following compound?

    a. I
    b. II
    c. III
    d. b and c
    e. a, b and c

I. $\xrightarrow{CH_3I}$

II. $\xrightarrow{LDA/THF}$ $\xrightarrow{CH_3I}$

III. $\xrightarrow{CH_3I}$ $\xrightarrow{H_3O^+}$

Answer: d   Section: 8

15. What is the <u>major</u> organic product from the following reaction?

   a. I
   b. II
   c. III
   d. IV
   e. V

I.

II.

III.

IV.

V.

Answer: e   Section: 10

16. Which of the following aldehydes does <u>not</u> undergo an aldol addition when mixed with base?

    a. I
    b. II
    c. III
    d. IV
    e. V

I. H—C—H (with =O above C)

II. (benzene ring)—CH$_2$CH (with =O above CH)

III. CH$_3$CH (with =O above CH)

IV. CH$_3$CH$_2$CH (with =O above CH)

V. (cyclohexane ring)—CH (with =O above CH)

Answer: a   Section: 11

17. Which of the following statements describes the first step in the mechanism of the aldol condensation?

    a. An alpha hydrogen is abstracted by the base to form an enolate anion.
    b. A nucleophilic base attacks the carbonyl carbon atom.
    c. The carbonyl oxygen is protonated by the base ion.
    d. The alpha hydrogen is abstracted by an acid to the enolate anion.
    e. The carbonyl oxygen of one aldehyde attacks the carbonyl carbon of another.

Answer: a   Section: 11

18. What is the major organic product of the following aldol addition?

    a. I
    b. II
    c. III
    d. IV
    e. V

Answer: b Section: 11

19. What is the major organic product of the following addition?

    a. I
    b. II
    c. III
    d. IV
    e. V

$$CH_3CH_2\overset{\overset{O}{\|}}{C}H \ + \ H-\overset{\overset{O}{\|}}{C}-H \ \underset{(excess)}{} \xrightarrow{\ ^-OH\ } \ ?$$

I. $CH_3CH_2\underset{\underset{OH}{|}}{C}H\overset{\overset{O}{\|}}{C}H$

II. $CH_3\underset{\underset{CH_3}{|}}{\overset{\overset{OH}{|}}{C}}-\overset{\overset{O}{\|}}{C}H$

III. $HOCH_2CH_2CH_2\overset{\overset{O}{\|}}{C}H$

IV. $CH_3\underset{\underset{CH_2OH}{|}}{C}H\overset{\overset{O}{\|}}{C}H$

V.   (cyclopentanone with OH on adjacent carbon)

Answer: d Section: 13

20. What is the <u>major</u> product of the following condensation?

a. I
b. II
c. III
d. IV
e. V

$$2\ CH_3CH_2\overset{\overset{\displaystyle O}{\|}}{C}OCH_2CH_3 \xrightarrow[2\cdot H_3O^+]{1\cdot CH_3CH_2O^-}\ ?$$

I. $CH_3CH_2\overset{\overset{\displaystyle OH}{|}}{\underset{\underset{\displaystyle CH_3}{|}}{C}}H\overset{\overset{\displaystyle O}{\|}}{C}HCOCH_2CH_3$

II.

III. $CH_3CH_2\overset{\overset{\displaystyle OH}{|}}{C}HCH_2CH_2\overset{\overset{\displaystyle O}{\|}}{C}OCH_2CH_3$

IV. $CH_3CH_2\overset{\overset{\displaystyle O}{\|}}{C}CH_2CH_2\overset{\overset{\displaystyle O}{\|}}{C}OCH_2CH_3$

V. $CH_3CH_2\overset{\overset{\displaystyle O}{\|}}{C}\overset{\overset{\displaystyle O}{\|}}{\underset{\underset{\displaystyle CH_3}{|}}{C}H}COCH_2CH_3$

Answer: e Section: 14

21. What materials are needed to prepare the following product?

    a. I
    b. II
    c. III
    d. IV
    e. V

I.   + CH₃CCl —H₃O⁺→

II.   + CH₃COR 1·CH₃CH₂O⁻ / 2· H₃O⁺

III.   + CH₃CH 1·CH₃CH₂O⁻ / 2· H₃O⁺

IV.   + CH₃CCH₃ 1·CH₃CH₂O⁻ / 2· H₃O⁺

V.   + CH₃COR 1·CH₃CH₂O⁻ / 2· H₃O⁺

Answer: b Section: 15

455

22. What is the <u>major</u> organic product of the following reaction?

    a. I
    b. II
    c. III
    d. IV
    e. V

I.

II.

III.

IV.

V.

Answer: a   Section: 16

23. What is the major organic product of the following reaction?

    a. I
    b. II
    c. III
    d. IV
    e. V

I.

II.

III.

IV.

V.

Answer: d Section: 16

24. What materials would you use to prepare the following compound using a
Robinson Annulation?

a. I
b. II
c. III
d. IV
e. V

I.

II.

III.

IV.

V.

Answer: c Section: 16

25. What is the <u>major</u> organic product of the following reaction?

$$CH_3CH_2\overset{\overset{\displaystyle O}{\|}}{C}OH \xrightarrow[\text{2. } Br_2, \text{ heat}]{\text{1. } Ag^+} ?$$

a. CH_3CH_2Br

b.
$$CH_3CH_2\overset{\overset{\displaystyle O}{\|}}{C}O\overset{\ominus\oplus}{Ag}$$

c.
$$CH_3CH_2\overset{\overset{\displaystyle O}{\|}}{C}Br$$

d.
$$CH_3CH_2\overset{\overset{\displaystyle O}{\|}}{C}OBr$$

e.
$$CH_3\overset{}{\underset{\overset{\displaystyle |}{Br}}{C}}H\overset{\overset{\displaystyle O}{\|}}{C}OH$$

Answer: a   Section: 17

26. What alkyl bromide should be used in the malonic ester synthesis of the following carboxylic acid?

    a. I
    b. II
    c. III
    d. IV
    e. V

$$CH_2CH_2\overset{\displaystyle O}{\overset{\displaystyle \|}{C}}OH$$

I. CH_3Br

II. CH_3CH_2Br

III. CH_2CH_2Br

IV. Br

V. CH_2Br

Answer: e   Section: 18

27. What alkyl bromide should be used in the acetoacetic ester synthesis of the following methyl ketone?

    a. I
    b. II
    c. III
    d. IV
    e. V

$$\text{Ph}-CH_2CH_2CH_2\overset{\overset{\displaystyle O}{||}}{C}CH_3$$

I. CH_3Br

II. CH_3CH_2Br

III.   Ph—CH_2CH_2Br

IV.   Ph—Br

V.   Ph—CH_2Br

Answer: c Section: 19

28. Which of the following compounds has the most stable enol tautomer?

    a. $CH_3\overset{\overset{\displaystyle O}{||}}{CH}$

    b. $CH_3\overset{\overset{\displaystyle O}{||}}{C}CH_3$

    c. $CH_3\overset{\overset{\displaystyle O}{||}}{C}CH_2\overset{\overset{\displaystyle O}{||}}{C}CH_3$

    d. $CH_3\overset{\overset{\displaystyle O}{||}}{C}-\overset{\overset{\displaystyle O}{||}}{C}CH_3$

    e. $CH_3\overset{\overset{\displaystyle O}{||}}{C}-\overset{\overset{\displaystyle O}{||}}{C}OH$

Answer: c Section: 2

461

SHORT ANSWER

29. Which of the following keto-enol tautomers is <u>more</u> stable? Explain.

| enol tautomer | keto tautomer |

Answer:
The enol form is more stable than the keto form because the enol form is aromatic but the keto form is not.

Section: 2

30. Explain why the following acid cannot be prepared by the malonic ester synthesis:

Answer:
Since bromobenzene is needed to prepare the acid, the reaction would not occur. This is true because it is difficult for a bromine to be a good leaving group if it is directly attached to a benzene ring (see S_N1/S_N2 reactions).

Section: 18

31. a) The based-catalyzed reaction of 1 mole of acetone with 2 moles of benzaldehyde yields a yellow solid, $C_{17}H_{14}O$. What is the structure of the product?
    b) What will be the result when only 1 mole of benzaldehyde is added?
    c) What will be the result when acetone is not added at all?
    d) What will be the result when benzaldehyde is not added at all?
    e) Why is the self-condensation of acetone unlikely in the presence of benzaldehyde?

Answer:
a) This is an aldol addition. (see graph a)
b) Using only 1 mole of benzaldehyde, the product would be the following. (see graph b)
c) Using no acetone, no reaction will occur since benzaldehyde has no α-hydrogens.
d) Using no benzaldehyde, the acetone will form an aldol addition product with the following structure. (see graph c)
e) This is because the benzaldehyde carbonyl carbon is more reactive than that of acetone toward nucleophilic attack. This is due to the electron releasing methyl groups on acetone. Also, benzaldehyde is less sterically hindered because it has a hydrogen on one side.

a)

$$\text{C}_6\text{H}_5-\text{CH}=\text{CHCCH}=\text{CH}-\text{C}_6\text{H}_5$$

with C=O above central carbon

b)

$$\text{C}_6\text{H}_5-\text{CH}=\text{CHCCH}_3$$

with C=O

c)

$$\underset{\overset{\displaystyle |}{\text{CH}_3}}{\text{CH}_3\overset{\overset{\displaystyle \text{OH}}{|}}{\text{C}}\text{CH}_2\overset{\overset{\displaystyle \text{O}}{\|}}{\text{C}}\text{CH}_3}$$

Section: 13

32. Propose a stepwise mechanism that would explain the following conversion:

   Answer:

   Section: 15

33. Show how an enolate can add to a carbonyl.

   Answer:

   Section: 11

34. Provide a detailed, stepwise mechanism for the base-catalyzed enolization of acetaldehyde.

   Answer:

   Section: 2

35. Provide a detailed, stepwise mechanism for the acid-catalyzed enolization of acetaldehyde.

Answer:

Section: 2

MULTIPLE CHOICE

36. Which of the following reagents will quantitatively convert an enolizable ketone to its enolate salt?

   a. lithium hydroxide
   b. lithium diisopropylamide
   c. methyllithium
   d. diethylamine
   e. pyridine

   Answer: b   Section: 7

SHORT ANSWER

37. Provide a detailed, stepwise mechanism for the α-bromination of acetone in base.

Answer:

Section: 4

38. Provide a detailed, stepwise mechanism for the formation of acetate and bromodiiodomethane from bromoacetone, hydroxide and iodine.

Answer:

Section: 4

MULTIPLE CHOICE

39. What is the carbon nucleophile which attacks molecular bromine in the acid-catalyzed α-bromination of a ketone?

a. an enolate
b. a Grignard reagent
c. an acetylide
d. a carbocation
e. an enol

Answer: e   Section: 4

40. The Hell-Volhard-Zelinsky reaction involves:

a. the α-bromination of carboxylic acids.
b. the α-bromination of ketones.
c. the bromination of alcohols.
d. the oxidation of aldehydes to acids.
e. none of the above.

Answer: a   Section: 5

41. Which of the following will alkylate a lithium enolate most rapidly?

    a. methyl bromide
    b. isopropyl bromide
    c. neopentyl bromide
    d. bromobenzene
    e. 2-methylbromobenzene

Answer: a   Section: 7

SHORT ANSWER

42. What iminium salt is produced in the reaction shown below?

Answer:

Section: 8

MULTIPLE CHOICE

43. An enolate attacks an aldehyde and the resulting product is subsequently protonated. What type of reaction is this?

    a. a Fischer esterification
    b. an acid-catalyzed aldol condensation
    c. a base-mediated aldol condensation
    d. a Hell-Volhard-Zelinsky reaction
    e. a Selman-Jones reaction

Answer: c   Section: 11

44. Which of the following could result from the dehydration of an aldol?

    a. 4-methyl-3-penten-2-one
    b. 4-methyl-4-penten-2-one
    c. 4-methyl-5-hexen-2-one
    d. 4-methyl-4-hexen-2-one
    e. 3-methyl-4-penten-2-one

    Answer: a Section: 12

45. In theory a poorly planned crossed aldol reaction can produce how many different aldol regioisomers?

    a. 1
    b. 2
    c. 3
    d. 4
    e. 5

    Answer: d Section: 13

SHORT ANSWER

46. Provide the sequence of steps necessary to synthesize the compound shown below from cyclohexene.

    Answer:
    1. O_3
    2. $S(CH_3)_2$
    3. NaOH, Δ
    Section: 16

47. What two molecules were condensed in an aldol reaction to produce $(CH_3)_3CCH{=}CHCOCH_3$?

    Answer: $(CH_3)_3CCHO$ and CH_3COCH_3 Section: 13

48. What two molecules were condensed in an aldol reaction to produce PhCH$=$CHCOPh?

    Answer: PhCHO and PhCOCH$_3$ Section: 13

49. Provide the structure of the Claisen product in the self-condensation of methyl phenylacetate.

Answer:

Section: 14

50. Starting with cyclohexene and employing a Dieckmann cyclization show how the compound below can be prepared.

Answer:
1. $KMnO_4$, ^-OH, Δ
2. H^+
3. CH_2N_2 <u>or</u> $SOCl_2$ followed by CH_3OH
4. $NaOCH_3$

Section: 16

51. What two starting materials yield $OHCCH_2CO_2CH_2CH_3$ as the mixed Claisen condensation product?

Answer: $HCO_2CH_2CH_3$ and $CH_3CO_2CH_2CH_3$ Section: 15

52. What product is formed in the mixed Claisen condensation between methyl benzoate and cyclohexanone?

Answer:

Section: 15

469

53. What product results when malonic ester is treated with the following sequence of reagents:
    1. $NaOCH_2CH_3$;
    2. $PhCH_2Br$;
    3. H_3O+, Δ?

    Answer: $PhCH_2CH_2CO_2H$ Section: 18

54. When compound X is heated, $PhCOCH(CH_3)_2$ and CO_2 are produced. Offer a structure for compound X.

    Answer: $PhCOC(CH_3)_2CO_2H$ Section: 17

MULTIPLE CHOICE

55. In the Michael reaction, addition to the α,β-unsaturated carbonyl occurs in a:

    a. 1,2-fashion.
    b. 1,3-fashion.
    c. 1,4-fashion.
    d. 1,5-fashion.
    e. Diels-Alder reaction.

    Answer: c   Section: 10

SHORT ANSWER

56. Provide a synthesis of the compound shown below from 2-cyclopenten-1-one and acetoacetic ester.

    Answer:
    From acetoacetic ester:
    1. NaOEt, EtOH
    2. 2-cyclopenten-1-one
    3. H_3O^+, Δ
    Section: 19

470

57. Provide the structure of the major organic product which results when PhCO$_2$CH$_2$CH$_3$ and CH$_3$CH$_2$CO$_2$CH$_2$CH$_3$ are heated in the presence of sodium ethoxide and the compound generated is subsequently treated with cold, dilute acid.

Answer:

Section: 15

58. Provide a detailed, stepwise mechanism for the transformation shown below.

Answer:

Section: 16

59. Why is the pKa of acetone 20 while that of ethane is 50? Explain the origin for this difference in acidity.

Answer:
When a proton is removed from ethane, the electrons left behind reside on a carbon atom. Because carbon is not very electronegative and there is no other effective way of delocalizing the negative charge, this carbanion is very unstable and very basic. When a proton is removed from acetone, a resonance delocalized enolate ion is formed. This enolate is stabilized through the delocalization of electron density onto an electronegative oxygen atom, thus the enolate is a much weaker base than the conjugate base of ethane. The weaker the conjugate base, the stronger the corresponding acid and the lower the pKa of this acid.

Section: 1

60. List the following carbonyl compounds in order of decreasing pKa: 2,4-pentanedione, diethyl malonate, propanal, ethyl acetate.

Answer: ethyl acetate > propanal > diethyl malonate > 2,4-pentanedione

Section: 1

MULTIPLE CHOICE

61. Which of the following routes works best for the preparation of 3-ethylcyclohexanone from cyclohexanone?

a. 1. LDA 2. Br_2 3. CH_3CH_2MgBr
b. 1. Br_2, H_3O^+ 2. $KOC(CH_3)_3$ 3. $(CH_3CH_2)_2CuLi$ 4. H_3O^+
c. 1. Br_2, H_3O^+ 2. CH_3CH_2OH, Δ
d. 1. NBS 2. Mg, ether 3. CH_3CHO
e. 1. Br_2, hʋ 2. NaOH 3. $NaOCH_2CH_3$ 4. H_3O^+

Answer: b   Section: 6

SHORT ANSWER

62. Enolates are frequently formed by treating an aldehyde or ketone with LDA. However, the LDA must first be generated by reacting diisopropylamine with butyllithium. Why don't chemists just use butyllithium itself to deprotonate the aldehyde or ketone and form the enolate?

Answer:
Butyllithium is a strong enough base but is too nucleophilic to be used to deprotonate carbonyl compounds. The butyllithium would attack the carbonyl carbon and form a covalent bond instead to deprotonating α to the carbonyl.

Section: 7

MULTIPLE CHOICE

63. A student proposed the following synthesis. Why did it fail?

   a. The use of LDA results in the thermodynamic product instead of the desired kinetic product.
   b. LDA is insoluble in the THF solvent so the reaction is too slow to occur at a useful rate.
   c. LDA acts as a nucleophile instead of a base in its reaction with cyclohexanone.
   d. The tertiary bromide is too sterically hindered to be attacked by the enolate.
   e. Instead of 2-methyl-2-bromopentane, 2-bromo-3-methylpentane should have been used to anticipate a cationic rearrangement that would occur.

Answer: d   Section: 8

64. What is the molecular formula of the major organic product of the following reaction sequence?

   a. $C_7H_{12}O$
   b. $C_7H_{14}O$
   c. $C_{11}H_{19}N$
   d. $C_{11}H_{21}N$
   e. $C_{11}H_{23}N$

Answer: a   Section: 9

65. When the 1,7-diester $CH_3O_2CCH_2CH_2CH(CH_3)CH_2CH_2CO_2CH_3$ is treated with sodium methoxide and the reaction mixture is subsequently neutralized with acid, what kind of compound is the major organic product?

    a. an acyclic β-diketone
    b. an acyclic β-ketoacid
    c. an acyclic β-diester
    d. a cyclic β-ketoester
    e. a cyclic β-diketone

    Answer: d   Section: 16

66. Which of the following compounds will react with methyl vinyl ketone in a Robinson annulation to generate the cyclic enone below?

    a. 1-pentene
    b. cyclohexanone
    c. 2-pentanone
    d. pentanal
    e. none of the above

    $CH_2CH_2CH_3$

    Answer: d   Section: 16

SHORT ANSWER

67. Provide a synthesis of 2-ethylhexanol using butanal as the only source of carbon.

    Answer:
    Depending on the conditions employed, hydrogenation could also reduce the carbonyl. This would render steps 4 and 5 unnecessary.

    $CH_3CH_2CH_2CHO$ →
    1. HO^- (aldol)
    2. H^+, Δ (dehydration)
    3. H_2, Pt (reduction of alkene)
    4. $NaBH_4$ (reduction of carbonyl)
    5. H_2O

    OH

    Section: 11

68. Provide the major organic product of the following reaction.

Answer:

Section: 15

Chapter 20: Carbohydrates

MULTIPLE CHOICE

1. Which of the following represents the general formula of a carbohydrate?

    a. C_nH_{2n+2}
    b. C_nH_{2n}
    c. C_nH_{2n-2}
    d. $C_n(H_2O)_n$
    e. $C_n(HO)_n$

    Answer: d Section: 98

2. Which of the following best represents the process of photosynthesis?

    a. disproportionation
    b. oxidation/reduction
    c. exothermic
    d. exergonic
    e. elimination

    Answer: b Section: 98

3. A carbohydrate composed of three to ten sugar molecules is called a(n):

    a. single carbohydrate
    b. disaccharide
    c. oligosaccharide
    d. polysaccharide
    e. monosaccharide

    Answer: c Section: 1

4. Which of the following best describes the sugar D-galactose,

    a. D-Aldohexose
    b. D-Ketohexose
    c. L-Aldohexose
    d. L-Ketohexose
    e. D-Aldopentose

    Answer: a Section: 1

5. Which of the following compounds is a D-sugar?

a.   CH₃C=O
     H— — OH
     H— — OH
     HO— — H
         CH₂OH

b.   O=CH
     H— — OH
     HO— — H
         CH₂OH

c.   O=CH
     HO— — H
     HO— — H
         CH₂OH

d.   CH₂OH
     HO— — H
     O= CH

e.   CH₂OH
     H— — OH
     O=CH

Answer: d   Section: 2

6. What is the relationship between the following compounds?

D-erythrose   L-erythrose

a. conformational isomers
b. constitutional isomers
c. identical
d. enantiomers
e. diastereomers

Answer: d   Section: 3

7. Which of the following corresponds to the definition of an aldopentose?
    I.   It is a monosaccharide.
    II. It contains a CHO group.
    III. It is a disaccharide.
    IV. It is an oligosaccharide.

    a. I. and II.
    b. II. and III.
    c. I. and IV.
    d. I. and III.
    e. I., II., and III.

    Answer: a Section: 1

8. At which carbon are the following sugars epimers of each other?

    a. C-1
    b. C-2
    c. C-3
    d. C-4
    e. C-5

    Answer: b Section: 3

9. At which carbon are the following sugars epimers of each other?

    a. C-1
    b. C-2
    c. C-3
    d. C-4
    e. C-5

    Answer: d Section: 3

10. At which carbon are the following sugars epimers of each other?

D-Fructose   D-Psicose

  a. C-1
  b. C-2
  c. C-3
  d. C-4
  e. C-5

Answer: c Section: 4

11. How many chirality centers are there in an aldohexose?

  a. 2
  b. 3
  c. 4
  d. 5
  e. 6

Answer: c Section: 3

12. How many chirality centers are there in a 2-ketohexose?

  a. 2
  b. 3
  c. 4
  d. 5
  e. 6

Answer: b Section: 4

13. How many stereoisomers are possible for a 2-ketohexose?

  a. 2
  b. 4
  c. 8
  d. 16
  e. 32

Answer: c Section: 4

14. Which of the following is/are the product(s) obtained from the reduction of D-fructose?

a.

D-Glucitol

b.

D-Mannitol

c.

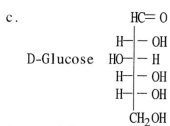

D-Glucose

d. a and b
e. a and c

Answer: d   Section: 5

15. Which of the following monosaccharides will form D-glucaric acid upon oxidation with nitric acid?

$$
\begin{array}{c}
\text{COOH} \\
\text{H}-\!\!-\text{OH} \\
\text{HO}-\!\!-\text{H} \\
\text{H}-\!\!-\text{OH} \\
\text{H}-\!\!-\text{OH} \\
\text{COOH}
\end{array}
$$

a. D-Glucose

$$
\begin{array}{c}
\text{HC}\!=\!\text{O} \\
\text{H}-\!\!-\text{OH} \\
\text{HO}-\!\!-\text{H} \\
\text{H}-\!\!-\text{OH} \\
\text{H}-\!\!-\text{OH} \\
\text{CH}_2\text{OH}
\end{array}
$$

b. L-Glucose

$$
\begin{array}{c}
\text{CH}_2\text{OH} \\
\text{H}-\!\!-\text{OH} \\
\text{H}-\!\!-\text{OH} \\
\text{HO}-\!\!-\text{H} \\
\text{H}-\!\!-\text{OH} \\
\text{HC}\!=\!\text{O}
\end{array}
$$

c. D-Glucitol

$$
\begin{array}{c}
\text{CH}_2\text{OH} \\
\text{H}-\!\!-\text{OH} \\
\text{HO}-\!\!-\text{H} \\
\text{H}-\!\!-\text{OH} \\
\text{H}-\!\!-\text{OH} \\
\text{CH}_2\text{OH}
\end{array}
$$

d. a and b
e. a and c

Answer: e Section: 5

16. Which of the following pairs of monosaccharides will form the _same_ osazone when reacted with phenylhydrazine?

a. D-Glucose and L-Glucose
b. D-Glucose and D-Gulose
c. D-Glucose and D-Fructose
d. D-Glucose and D-Allose
e. D-Glucose and D-Galactose

Answer: c Section: 6

17. Which monosaccharides are formed in the Kiliani-Fischer synthesis starting with D-xylose?

a. D-Glucose and D-Mannose
b. D-Gulose and D-Idose
c. D-Galactose and D-Talose
d. D-Allose and D-Altrose
e. D-Ribose and D-Arabinose

Answer: b Section: 7

18. Which two monosaccharides can be degraded by the Ruff Degradation to D-arabinose?

$$
\begin{array}{c}
HC{=}O \\
HO{-}|{-}H \\
H{-}|{-}OH \\
H{-}|{-}OH \\
CH_2OH
\end{array}
$$

a. D-Allose and D-Altrose
b. D-Gulose and D-Idose
c. D-Galactose and D-Talose
d. D-Erythrose and D-Threose
e. D-Glucose and D-Mannose

Answer: e Section: 8

19. An aqueous solution of glucose behaves as an aldehyde because:

a. it is hydrolyzed by water to the free aldehyde.
b. it is a ketone, but is in equilibrium with the aldehyde form.
c. glucose is actually a cyclic aldehyde.
d. its cyclic hemiacetal, the predominant form, is in equilibrium with the free aldehyde form.
e. it can be oxidized with periodic acid.

Answer: d Section: 10

20. C-2 epimeric aldohexoses give the same:

    a. osazone.
    b. aldaric acid when oxidized with HNO_3.
    c. products with the Kiliani-Fischer synthesis.
    d. a and b.
    e. b and c.

    Answer: a   Section: 6

21. Which of the following statements best describes the meaning of <u>mutorotation</u>?

    a. A rapid exchange between the α and β forms of diasternomeric sugars
    b. A rapid exchange between the D and L forms
    c. A slow exchange between hydrogen and deuterated hydrogen
    d. A slow change in optical rotation to reach an equilibrium value
    e. A slow change in absolute configuration to reach an equilibrium value

    Answer: d   Section: 10

22. Which of the following is the cyclic hemiacetal formed from 4-hydroxyheptanal?

    a. I
    b. II
    c. III
    d. IV
    e. V

    I.

    II.

    III.

    IV.

    V.

    Answer: c   Section: 10

23. Which of the following is the Haworth projection of β-D-galactose?

$$\begin{array}{c}
HO{-}\!-\!H \\
H{-}\!-\!OH \\
HO{-}\!-\!H\ O \\
HO{-}\!-\!H \\
H{-} \\
CH_2OH
\end{array}$$

   a. I
   b. II
   c. III
   d. IV
   e. V

I.

II.

III.

IV.

V.

Answer: a Section: 10

24. Which of the following statements best describes the meaning of a glycoside?

   a. It is the mirror image of a sugar.
   b. It is the hemiacetal of a sugar.
   c. It is the acetal of a sugar.
   d. It is the enantiomer of a sugar.
   e. It is the enol-keto form of a sugar.

Answer: c Section: 13

25. Which of the following is methyl-α-D-glucoside?

    a. I
    b. II
    c. III
    d. IV
    e. V

I.

II.

III.

IV.

V.

Answer: a Section: 13

26. Which of the following compounds is a nonreducing sugar?

    a. I
    b. II
    c. III
    d. IV
    e. V

Answer: e Section: 13

27. Which of the following statements best describes the difference between amylose and amylopectin?

    a. Amylose is a branched polysaccharide while amylopectin is a chain polysaccharide.
    b. Amylose is a straight-chain polysaccharide while amylopectin is a branched polysaccharide.
    c. Amylose contains α-1,6-glycosidic linkage which amylopectin does not contain.
    d. Amylose is composed of thousands of D-glucose units while amylopectin is composed of thousands of D-galactose units.
    e. Amylose is one of the largest molecules found in nature while amylopectin is one of the smallest molecules found in nature.

Answer: b Section: 4

28. Which of the following statements best describes the main structural difference between cellulose and chitin?

    a. The C-2 carbon in chitin contains an N-acetyl amino group instead of an OH group.
    b. The C-2 carbon in cellulose contains an N-acetyl amino group instead of an OH group.
    c. Chitin is a β-1,4-glycosidic linkage while cellulose is a α-1,4-glycosidic linkage.
    d. Chitin is a α-1,4-glycosidic linkage while cellulose is a α-1,4-glycosidic linkage.
    e. Chitin is a branched polysaccharide while cellulose is a chain polysaccharide.

Answer: a   Section: 18

29. Which of the following is the Haworth projection of a α-D-tagatose?

    a. I
    b. II
    c. III
    d. IV
    e. V

Answer: e   Section: 10

30. Which of the following blood types is known as the universal donor?

    a. A
    b. B
    c. AB
    d. O
    e. ABO

    Answer: d Section: 20

31. Which of the following blood types is known as the universal acceptor?

    a. A
    b. B
    c. AB
    d. O
    e. ABO

    Answer: c Section: 20

32. Which of the following is true about sucrose?

    a. It hydrolyzes to fructose and glucose.
    b. It is a reducing sugar.
    c. It is a monosaccharide.
    d. It undergoes mutorotation in water.

    Answer: a Section: 17

SHORT ANSWER

33. Is the following structure of glucose a D or L configuration? Explain.

    Answer:
    This is L-glucose because the OH group attached to the bottom-most
    stereogenic center is on the left.
    Section: 2

34. When D-tagatose is added to a basic aqueous solution, an equilibrium mixture of three monosaccharides is obtained. Identify the monosaccharides.

$$
\begin{array}{c}
CH_2OH \\
| \\
C=O \\
| \\
HO-C-H \\
| \\
HO-C-H \\
| \\
H-C-OH \\
| \\
CH_2OH
\end{array}
$$

D-tagatose

Answer:

$$
\begin{array}{cccc}
CH_2OH & CHOH & H-C=O & H-C=O \\
| & \| & | & | \\
C=O & C-OH & HO-C-H & H-C-OH \\
| & | & | & | \\
HO-C-H \quad ^{\ominus}OH & HO-C-H \quad ^{\ominus}OH & HO-C-H & HO-C-H \\
| & | & | & | \\
HO-C-H \quad \rightleftharpoons & HO-C-H \quad \rightleftharpoons & HO-C-H & HO-C-H \\
| & | & | & | \\
H-C-OH \quad H_2O & H-C-OH \quad H_2O & H-|-OH & H-|-OH \\
| & | & | & | \\
CH_2OH & CH_2OH & CH_2OH & CH_2OH
\end{array}
$$

D-tagatose   enol form of   D-talose   D-galactose
            D-tagatose

Therefore the three monosaccharides are: D-tagatose, D-galactose, and D-talose.

Section: 5

35. Why is β-D-glucopyranose more stable in nature than α-D-glucopyranose?

Answer:
The five hydroxy groups in β-D-glucopyranose are located on the equatorial position of the chair conformation. This arrangement is more stable than the chair conformation in which one of the hydroxy groups of -D-glucopyranose is in the axial position.

Section: 11

36. What is the main structural difference between amylose and cellulose?

Answer:
The main difference is that amylose is composed of glucose units joined by α-1,4-glycosidic linkages while cellulose is composed of glucose units joined by β-1,4-glycosidic linkages.

Section: 18

MULTIPLE CHOICE

37. All chiral D-sugars rotate plane-polarized light:

    a. clockwise.
    b. counterclockwise.
    c. +20.0° .
    d. in a direction that cannot be predicted but must be determined experimentally.
    e. since they are optically inactive.

Answer: d   Section: 2

SHORT ANSWER

38. Draw the Fischer projection for the open-chain form of D-glucose.

Answer:

Section: 3

39. Draw the Fischer projection for the open-chain form of D-fructose.

Answer:

Section: 4

40. Draw the Fischer projection for the open-chain form of D-erythrose.

Answer:

$$\begin{array}{c} \text{CHO} \\ \text{H} \!-\!\!\!-\!\!\!- \text{OH} \\ \text{H} \!-\!\!\!-\!\!\!- \text{OH} \\ \text{CH}_2\text{OH} \end{array}$$

Section: 3

MULTIPLE CHOICE

41. In solution, glucose exists as:

    a. the open-chain form only.
    b. the cyclic hemiacetal form only.
    c. the cyclic acetal form only.
    d. an equilbrium mixture of the open-chain form and cyclic acetal forms.
    e. an equilbrium mixture of the open-chain form and cyclic hemiacetal forms.

    Answer: e   Section: 10

42. When a monosaccharide reacts to give the pyranose form from its open-chain form, how many distinct pyranose forms are possible?

    a. 1
    b. 2
    c. 2^n, where n is the number of carbons present
    d. $4n + 2$, where n is the number of carbons present
    e. 4

    Answer: b   Section: 10

43. Six-membered cyclic hemiacetals and five-membered cyclic hemiacetals are called respectively:

    a. mannoses and xyloses.
    b. maltoses and arabinoses.
    c. pyranoses and fructoses.
    d. glyoses and fructoses.
    e. none of the above.

    Answer: e   Section: 10

44. Anomers of D-glucopyranose differ in their stereochemistry at:

    a. C1.
    b. C2.
    c. C3.
    d. C4.
    e. C5.

    Answer: a Section: 10

45. A pyranose with the hydroxyl group on the anomeric carbon pointing up in the Haworth structure is designated:

    a. a'.
    b. b'.
    c. α.
    d. β.
    e. *f.*

    Answer: d Section: 10

SHORT ANSWER

46. When pure α-D-glucopyranose is dissolved in water, the optical rotation of the resulting solution changes over a period of time. What is the name of this phenomenon and why does it occur?

    Answer:
    The phenomenon is called mutarotation. In solution, α-D-glucopyranose is converted to the equilbrium mixture of α and β anomers via the open-chain form.
    Section: 10

47. Draw the Haworth structure of α-D-ribofuranose

    Answer:

    Section: 10

48. Draw the Haworth structure of β-D-glucopyranose

Answer:

Section: 10

49. Draw the more stable chair conformer of α-D-glucopyranose.

Answer:

Section: 10

MULTIPLE CHOICE

50. Reduction of a 2-ketohexose with $NaBH_4$ yields:

a. a single aldohexose.
b. a mixture of acetals.
c. a mixture of alditols.
d. a mixture of cyclic hemiacetals.
e. a single pyranose.

Answer: c Section: 5

SHORT ANSWER

51. Provide the Fischer projection of the open-chain form of the aldonic acid which results when L-glucose is treated with bromine water.

Answer:

Section: 5

MULTIPLE CHOICE

52. When D-threose is treated with NaBH$_4$,:

    a. a 70:30 mixture of enantiomeric alditols results.
    b. a 50:50 mixture of enantiomeric alditols results.
    c. a meso alditol is produced.
    d. the product mixture contains two diastereomeric alditols.
    e. an optically active alditol is produced.

Answer: e Section: 5

SHORT ANSWER

53. An optically active D-aldopentose produced an optically inactive aldaric acid upon treatment with HNO$_3$. When this aldopentose was subjected to a Ruff degradation, a D-aldotetrose was generated. This aldotetrose gave an optically inactive alditol upon reduction with NaBH$_4$. Use these data to provide the structure of the starting D-aldopentose.

Answer:

```
        CHO
   H ——|—— OH
   H ——|—— OH
   H ——|—— OH
       CH₂OH
```

Section: 5

54. An optically active D-aldopentose produced an optically active alditol upon treatment with NaBH$_4$. When this aldopentose was subjected to a Ruff degradation, a D-aldotetrose was generated. This aldotetrose gave an optically active aldaric acid upon oxidation with HNO$_3$. Use these data to provide the structure of the starting D-aldopentose.

Answer:

```
        CHO
  HO ——|—— H
  HO ——|—— H
   H ——|—— OH
       CH₂OH
```

Section: 5

55. An optically active D-aldopentose produced an optically inactive alditol upon treatment with $NaBH_4$. When this aldopentose was subjected to a Ruff degradation, a D-aldotetrose was generated. This aldotetrose gave an optically active aldaric acid upon oxidation with HNO_3. Use these data to provide the structure of the starting D-aldopentose.

Answer:

```
      CHO
      |
 H—  —OH
HO—  —H
 H—  —OH
      |
     CH2OH
```

Section: 5

56. Provide the structure of the product which results when D-ribose is treated with bromine water.

Answer:

```
     COOH
      |
 H—  —OH
 H—  —OH
 H—  —OH
      |
    CH2OH
```

Section: 5

MULTIPLE CHOICE

57. Which of the following would give a positive Tollen's test?

   a. α-D-glucopyranose
   b. methyl β-D-glucopyranoside
   c. sucrose
   d. methyl α-D-ribofuranoside
   e. none of the above

Answer: a   Section: 15

SHORT ANSWER

58. Under what conditions is the methyl glycoside of galactose prepared?

Answer: CH_3OH, H^+ (nonaqueous)   Section: 13

MULTIPLE CHOICE

59. Two D-aldopentoses give the same D-aldotetrose upon Ruff degradation. The two aldopentoses are:

    a. enantiomers.
    b. meso compounds.
    c. C2 epimers.
    d. C4 epimers.
    e. D-glucose and D-galactose.

    Answer: c Section: 8

SHORT ANSWER

60. What nucleophilic carbon species is used in the Kiliani-Fischer synthesis?

    Answer: cyanide Section: 7

61. An optically active D-aldopentose **A** produced an optically active alditol **B** upon treatment with sodium borohydride. When this aldopentose was subjected to a Ruff degradation, a D-aldotetrose **C** was generated. This aldotetrose yielded an optically inactive aldaric acid **D** upon oxidation with nitric acid. Use these data to provide the structures of compounds **A**, **B**, **C**, and **D**.

    Answer:

    Section: 5

MULTIPLE CHOICE

62. Which of the following sugars yields the same alditol upon reduction with NaBH4 as does D-glucose?

    a. L-glucose
    b. D-mannose
    c. L-gulose
    d. D-arabinose
    e. none of the above

    Answer: c   Section: 5

63. When D-sorbose (shown below) is treated with NaBH4, which of the following is produced?

    a. a single optically active alditol
    b. a single optically inactive alditol
    c. a racemic mixture of alditols
    d. an optically inactive mixture of epimeric alditols
    e. an optically active mixture of epimeric alditols

    Answer: e   Section: 5

SHORT ANSWER

64. Which of the D-aldopentoses yields an optically inactive aldonic acid upon oxidation with bromine water? Explain your answer.

    Answer:
    None of the aldonic acids produced upon bromine water oxidation of a D-aldopentose is optically inactive. The top of the Fischer projection is a -CO$_2$H group and the bottom is a -CH$_2$OH group. This precludes the existence of an internal mirror plane in any of these aldonic acids. Thus all of these aldonic acids will be chiral and optically active.

    Section: 5

65. Provide the structure of the osazone which results when D-idose is treated with 3 equivalents of phenyl hydrazine. Also draw the structures of two other hexoses which yield the same osazone upon treatment with phenylhydrazine.

Answer: See graph.

Section: 6

MULTIPLE CHOICE

66. Which of the following sequences is known as a Kiliani-Fischer chain elongation?

a. 1. Fe^{3+}, H_2O_2 2. Br_2, H_2O 3. $Ca(OH)_2$
b. 1. CN^- 2. H_2, $Pd/BaSO_4$ 3. H_3O^+
c. 1. $NaBH_4$ 2. H_2O 3. Br_2, H_2O
d. 1. acetic anhydride (excess), pyridine 2. CH_3I (excess), Ag_2O
e. 1. CH_3OH, H^+, Δ 2. $AgNO_3$, Δ

Answer: b   Section: 7

SHORT ANSWER

67. Provide the Haworth projection of ethyl-β-D-mannopyranoside.

Answer:

Section: 13

MULTIPLE CHOICE

68. When α-D-xylofuranose is treated with excess methyl iodide in the presence of Ag_2O, how many equivalents of methyl iodide react with each equivalent of the furanose?

  a. 0
  b. 1
  c. 4
  d. 5
  e. 6

Answer: d   Section: 12

SHORT ANSWER

69. Provide the structure of the more stable chair conformer which results when β-D-galactose is treated with excess acetic anhydride/pyridine.

Answer:

Section: 12

MULTIPLE CHOICE

70. Which of the following statements regarding methyl-β-L-glucopyranoside is correct?

    a. This glycoside is a reducing sugar.
    b. This glycoside undergoes mutorotation in aqueous base.
    c. This glycoside will undergo no reaction when treated with excess CH_3I/Ag_2O.
    d. This glycoside will be hydrolyzed to the cyclic hemiacetal in dilute aqueous acid.
    e. This glycoside will readily undergo Kiliani-Fischer chain elongation.

    Answer: d   Section: 13

Chapter 21:   Amino Acids, Peptides, and Proteins

MULTIPLE CHOICE

1. The monomeric units that make up peptides and protein polymers are:

    a. Nucleic acids
    b. Amino acids
    c. Oligosaccharides
    d. Amylopectins
    e. Celluloses

    Answer: b Section: 98

2. Which labeled bond in the following molecule is known as the peptide bond?

    a. 1
    b. 2
    c. 3
    d. 4
    e. 5

    Answer: c Section: 7

3. What are enzymes?

    a. Saccharides that catalyze chemical reactions
    b. Nucleic acids that catalyze chemical reactions
    c. Unsaturated fats that catalyze chemical reactions
    d. DNA molecules that catalyze chemical reactions
    e. Proteins that catalyze chemical reactions

    Answer: e Section: 98

4. Which of the following molecules is the skeletal structure of an amino acid?

a.

$$R{-}C{=}CH{-}COO^{\ominus}$$
$$|$$
$$ONH_3$$

b.

$$R{-}CH{-}CH_2{-}COO^{\ominus}$$
$$|$$
$$\underset{\oplus}{NH_3}$$

c.

$$R{-}CH{-}COO^{\ominus}$$
$$|$$
$$\underset{\oplus}{NH_3}$$

d.

$$\overset{O}{\overset{||}{R{-}CH{-}C{-}COO^{\ominus}}}$$
$$|$$
$$\underset{\oplus}{NH_3}$$

e.

$$\overset{O}{\overset{||}{R{-}CH{-}C{-}NH_3^{\oplus}}}$$
$$\underset{COO^{\ominus}}{|}$$

Answer: c Section: 1

5. Which of the following amino acids has an aliphatic R group?

a. Serine
b. Cysteine
c. Asparagine
d. Tyrosine
e. Leucine

Answer: e Section: 1

6. Which of the following amino acids has an aromatic R group?

a. Serine
b. Cysteine
c. Asparagine
d. Tyrosine
e. Leucine

Answer: d Section: 1

7. Which of the following amino acids has a sulfur in the R group?

   a. Serine
   b. Cysteine
   c. Asparagine
   d. Tyrosine
   e. Leucine

   Answer: b   Section: 1

8. Which of the following amino acids has a heterocyclic R group?

   a. Glycine
   b. Threonine
   c. Proline
   d. Aspartic acid
   e. Arginine

   Answer: c   Section: 1

9. The α-carbon of all the amino acids is a chirality center except for:

   a. Glycine
   b. Threonine
   c. Proline
   d. Aspartic acid
   e. Arginine

   Answer: a   Section: 1

10. Which of the following is an L-amino acid?
    a. I
    b. II
    c. III
    d. IV
    e. V

I. $H_3\overset{\oplus}{N}$ ─┼─ H   (COO$^{\ominus}$ top, H bottom)

II. $H_3\overset{\oplus}{N}$ ─┼─ H   (COO$^{\ominus}$ top, R bottom)

III. H ─┼─ $\overset{\oplus}{N}H_3$ (COO$^{\ominus}$ top, R bottom)

IV. $H_3\overset{\oplus}{N}$ ─┼─ R   (COO$^{\ominus}$ top, H bottom)

V.   R ─┼─ $\overset{\oplus}{N}H_3$ (H top, COO$^{\ominus}$ bottom)

Answer: b Section: 2

11. Which of the following best represents the structure of an amino acid in basic solution (pH = 11)?

a.
   R─CH─COO$^{\ominus}$
       │
       $\overset{\oplus}{N}H_2$

b. R─CH─COOH
       │
       NH_2

c. R─CH─COOH
       │
       $\overset{\oplus}{N}H_3$

d.
   R─CH─COO$^{\ominus}$
       │
       $\overset{\oplus}{N}H_3$

e.
   R─CH─COO$^{\ominus}$
       │
       NH_2

Answer: e Section: 3

12. Which of the following is a <u>zwitterion</u>?

a.
$$R-CH-COO^{\ominus}$$
$$|$$
$$^{\oplus}NH_2$$

b. $R-CH-COOH$
$$|$$
$$NH_2$$

c. $R-CH-COOH$
$$|$$
$$^{\oplus}NH_3$$

d.
$$R-CH-COO^{\ominus}$$
$$|$$
$$^{\oplus}NH_3$$

e.
$$R-CH-COO^{\ominus}$$
$$|$$
$$NH_2$$

Answer: d Section: 3

13. Which of the following amino acids has a side chain with an ionizable proton and can exist in four different forms, depending on the pH of the solution?

a. I
b. II
c. III
d. IV
e. V

I. Proline

II. Serine $HO-CH_2-\overset{\overset{\displaystyle H}{|}}{\underset{\underset{\displaystyle \overset{\oplus}{NH_3}}{|}}{C}}-COO^{\ominus}$

III. Histidine

IV. Alanine $CH_3-\overset{\overset{\displaystyle H}{|}}{\underset{\underset{\displaystyle \overset{\oplus}{NH_3}}{|}}{C}}-COO^{\ominus}$

V. Glycine $H-\overset{\overset{\displaystyle H}{|}}{\underset{\underset{\displaystyle \overset{\oplus}{NH_3}}{|}}{C}}-COO^{\ominus}$

Answer: c Section: 3

14. What is the pI of glycine? The structure and pKa values are shown below.

$$H-\underset{\underset{\oplus}{NH_3}}{\overset{\overset{H}{|}}{\underset{|}{C}}}-COOH \longleftarrow pKa = 2.34$$

pKa = 9.60

a. 7.26
b. 5.97
c. 3.63
d. 7.50
e. 11.94

Answer: b Section: 4

15. What is the pI of arginine? The structure and pKa values are shown below.

$$NH_2-\underset{\underset{\oplus}{\underset{NH_2}{||}}}{C}-NH-(CH_2)_3-\underset{\underset{\oplus}{NH_3}}{\overset{\overset{H}{|}}{C}}-COOH \longleftarrow pKa = 2.17$$

pKa = 12.48   pKa = 9.04

a. 10.76
b. 7.90
c. 5.61
d. 7.33
e. 9.67

Answer: a Section: 5

16. What is <u>electrophoresis</u>?

    a. A technique that separates amino acids on the basis of their polarity
    b. A technique that separates amino acids on the basis of their solubility in water
    c. A technique that separates amino acids on the basis of their pI values
    d. A technique that separates amino acids on the basis of pKa of α-COOH values
    e. A technique that separates amino acids on the basis of pKa of α-$^+NH_3$ values

    Answer: c Section: 5

17. What is <u>thin-layer</u> <u>chromatography</u>?

    a. A technique that separates amino acids on the basis of their polarity
    b. A technique that separates amino acids on the basis of their solubility in water
    c. A technique that separates amino acids on the basis of their pI values
    d. A technique that separates amino acids on the basis of pKa of α-COOH values
    e. A technique that separates amino acids on the basis of pKa of α-$^+NH_3$ values

    Answer: a Section: 5

18. Which of the following amino acids will be closest to the origin when separated by thin-layer chromatography?

a. I
b. II
c. III
d. IV
e. V

I. Serine $HO-CH_2-\overset{\overset{\displaystyle H}{|}}{\underset{\underset{\displaystyle \oplus NH_3}{|}}{C}}-COO^{\ominus}$.

II. Alanine $CH_3-\overset{\overset{\displaystyle H}{|}}{\underset{\underset{\displaystyle \oplus NH_3}{|}}{C}}-COO^{\ominus}$

III. Cysteine $HS-CH_2-\overset{\overset{\displaystyle H}{|}}{\underset{\underset{\displaystyle \oplus NH_3}{|}}{C}}-COO^{\ominus}$

IV. Aspartate $^{\ominus}OOCCH_2-\underset{\underset{\displaystyle \oplus NH_3}{|}}{CH}COO^{\ominus}$

V. Phenylalanine $\langle\!\langle\bigcirc\rangle\!\rangle-CH_2-\overset{\overset{\displaystyle H}{|}}{\underset{\underset{\displaystyle \oplus NH_3}{|}}{C}}-COOH$

Answer: d   Section: 5

19. Which of the following amino acids will be retained longest in the column when separated by <u>cation-exchange</u> <u>chromatography</u>?

    a. I
    b. II
    c. III
    d. IV
    e. V

    I.   Serine $HO-CH_2-\overset{\overset{\textstyle H}{|}}{\underset{\underset{\textstyle \overset{\oplus}{NH_3}}{|}}{C}}-COO^{\ominus}$

    II. Alanine $CH_3-\overset{\overset{\textstyle H}{|}}{\underset{\underset{\textstyle \overset{\oplus}{NH_3}}{|}}{C}}-COO^{\ominus}$

    III. Cysteine $HS-CH_2-\overset{\overset{\textstyle H}{|}}{\underset{\underset{\textstyle \overset{\oplus}{NH_3}}{|}}{C}}-COO^{\ominus}$

    IV. Lysine $\overset{\oplus}{H_3}N-CH_2CH_2CH_2CH_2\underset{\underset{\textstyle \overset{\oplus}{NH_3}}{|}}{CH}-COOH$

    V.   Phenylalanine $\langle\!\!\bigcirc\!\!\rangle-CH_2-\overset{\overset{\textstyle H}{|}}{\underset{\underset{\textstyle \overset{\oplus}{NH_3}}{|}}{C}}-COOH$

Answer: d Section: 5

20. What is the <u>major</u> organic product of the following reaction?

$$CH_3-\overset{\overset{\text{O}}{\|}}{C}-OH \xrightarrow[\text{2. } H_3O^+]{\text{1. } Br_2/P \text{ excess}} \xrightarrow[NH_3]{} ?$$

a.
$$CH_3\overset{\overset{\text{O}}{\|}}{C}NH_2$$

b.
$$\overset{}{\underset{\overset{\|}{NH_3} \atop \oplus}{CH_2COO}}^{\ominus}$$

c.
$$\underset{\overset{|}{Br}}{CH_2\overset{\overset{\text{O}}{\|}}{C}NH_2}$$

d.
$$\underset{\overset{|}{NH_3} \atop \oplus}{CH_2\overset{\overset{\text{O}}{\|}}{C}NH_2}$$

e.
$$\underset{}{CH_3\overset{\overset{\text{Br}}{|}}{C}=NH}$$

Answer: b Section: 6

21. Which of the following is an acceptable name for the peptide sequence shown below?

Glu-Glu-His-Val-Cys

a. 1,2-Diglutamylhistidylvalylcysteine
b. Glutamylglutamylhistidylvalylcysteine
c. 1-Glutamyl-2-glutamylhistidylvalylcysteine
d. Diglutamylhistidylvalylcysteine
e. Biglutamylhistidylvalylcysteine

Answer: b Section: 6

22. What is the product obtained from the mild oxidation of the thiol shown below?

   2 R–SH $\xrightarrow[\text{oxidation}]{\text{mild}}$
   Thiol

   a. R-S-R
   b. R-OH
   c. R-S-S-R
   d. R-OO-R
   e. R-S-OH

   Answer: c Section: 7

23. Which of the following structures can polypeptides have?

   a. Primary structure
   b. Secondary structure
   c. Tertiary structure
   d. Quaternary structure
   e. All of the above

   Answer: e Section: 11

24. Which of the following is the first step in the determination of the primary structure of proteins?

   a. Determining the number and kind of amino acids in the peptide
   b. Reducing the disulfide bridges in the protein
   c. Protecting the N-terminal of the peptide
   d. Protecting the C-terminal of the peptide
   e. Hydrolyzing the protein with dilute acid

   Answer: b Section: 12

25. Phenyl isothiocyanate (PITC) is commonly known as Edman's reagent. Edman's reagent is widely used to identify which of the following?

   a. The number and kind of amino acids in the protein
   b. The disulfide bridges in the protein
   c. C-terminal amino acid of the protein
   d. N-terminal amino acid of the protein
   e. The number of peptide bonds in the protein

   Answer: d Section: 12

26. Carboxypeptidase is an enzyme used in the determination of the primary structure of a protein. Specifically, which of the following does it identify?

    a. The number and kind of amino acids in the protein
    b. The disulfide bridges in the protein
    c. C-terminal amino acid of the protein
    d. N-terminal amino acid of the protein
    e. The number of peptide bonds in the protein

Answer: c Section: 12

27. A Decapeptide undergoes partial hydrolysis to give peptides whose amino acids are shown below. Reaction of the decapeptide with Edman's reagent releases PTH-leu. What is the sequence of the decapeptide?

    (1) Pro, Ser   (5) Glu, Ser, Val, Pro
    (2) Gly, Gly   (6) Glu, Pro, Gly
    (3) Met, Ala, Leu   (7) Met, Leu
    (4) Gly, Ala   (8) His, Val

    a. Leu-Met-Ala-Gly-Gly-Glu-Pro-Ser-His-Val
    b. Leu-Met-Ala-Gly-Gly-Glu-Ser-Pro-Val-His
    c. Leu-Ala-Met-Gly-Gly-Glu-Pro-Ser-Val-His
    d. Leu-Met-Gly-Ala-Gly-Glu-Pro-Ser-Val-His
    e. Leu-Met-Ala-Gly-Gly-Glu-Pro-Ser-Val-His

Answer: e Section: 12

28. Which of the following may characterize the secondary structure of proteins?

    a. Conformation of the protein backbone
    b. α-Helix
    c. Parallel β-pleated sheet
    d. Anti-parallel β-pleated sheet
    e. All of the above

Answer: e Section: 13

29. Which of the following is the quaternary structure of proteins concerned with?

    a. Sequence of amino acids in the peptide chain
    b. Description of the way the peptide chains are arranged with respect to each other
    c. Location of the disulfide bridges in the peptide chain
    d. Conformation of the protein backbone
    e. Three-dimensional arrangement of all atoms in the protein

Answer: b Section: 15

30. Which of the following protein structures does <u>denaturation</u> destroy?

    a. Primary and secondary structures
    b. Secondary and tertiary structures
    c. Tertiary and quaternary structures
    d. Secondary, tertiary, and quaternary structures

    Answer: d Section: 16

31. Which of the following are capable of denaturing proteins?

    a. Organic solvents
    b. Detergents
    c. Extreme pH
    d. Heat
    e. All of the above

    Answer: e Section: 16

SHORT ANSWER

32. Explain what essential amino acids are and list all of them.

    Answer:
    An essential amino acid is one that humans must obtain from their diet because it cannot be synthesized at all or in sufficient quantity in the body. They are valine, leucine, isoleucine, threonine, methionine, lysine, arginine, phenylalanine, histidine, and tryptophan.
    Section: 1

33. In nature, which is the most commonly found isomer, L-amino acids or D-amino acids?

    Answer:
    Unlike monosaccharides where the D-isomers are more common, most amino acids found in nature have the L-configuration.
    Section: 2

34. Draw the form in which glutamate exists at pH = 0.

Answer:
At pH = 0, groups will be in their acidic form:

$$HO-\overset{\overset{\displaystyle O}{\|}}{C}-CH_2CH_2-\underset{\underset{\displaystyle \oplus}{\underset{\displaystyle NH_3}{|}}}{CH}-\overset{\overset{\displaystyle O}{\|}}{C}-OH$$

Section: 3

35. Define what is meant by isoelectric point (pI) and give an example.

Answer:
Isoelectric point is the pH of an amino acid at which it has no net electric charge. For amino acids that have no ionizable side chain, the pI value is the average of its two pKa's. If the amino acid has an ionizable side chain, the pI value is the average of the pKa's of similarly ionizable groups.

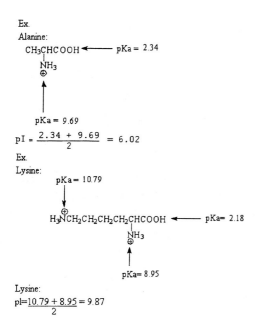

Ex.
Alanine:

pI $= \dfrac{2.34 + 9.69}{2} = 6.02$

Ex.
Lysine:

Lysine:
pI$=\dfrac{10.79 + 8.95}{2} = 9.87$

Section: 4

36. Rate the following amino acids in decreasing order of migration towards the cathode (negative electrode) when separated by electrophoresis in a solution of pH= 7.3.

    I. Lysine: pI= 9.87
    II. Alanine: pI= 6.02
    III. Aspartate: pI= 5.95

    Answer: I > II > III Section: 5

37. What is an <u>amino acid analyzer</u>?

    Answer:
    An instrument that automates ion-exchange chromatography. After the amino acids are separated, ninhydrin is added to each fraction. The concentration of the amino acid can be determined by measuring the absorbance at 570 nm. In this way, the identity and the relative amount of each amino acid in a protein can be determined.
    Section: 5

38. Explain what is meant by <u>kinetic resolution</u> and give an example.

    Answer:
    A separation technique in which an enzyme (such as pig kidney aminoacylase) is used to separate a racemic mixture of amino acids into its enantiomers. The separation depends on the rate of reaction of the enzyme with the two enantiomers.
    Section: 6

39. Explain the importance of the Merrifield method in peptide synthesis.

    Answer:
    In 1969, Bruce Merrifield developed a method that revolutionized the synthesis of peptides by producing them faster and in higher yields. This method became known as automated solid-phase peptide synthesis. The method won him the Nobel Prize in chemistry in 1984.
    Section: 10

40. Describe what a β-pleated sheet is and explain the difference between parallel and anti-parallel β-pleated sheets.

    Answer:
    A β-pleated sheet is a type of secondary structure of a protein in which the polypeptide backbone is extended in a zigzag structure. In parallel sheets, adjacent chains run in the same direction, whereas adjacent chains run in opposite directions in anti-parallel sheets.
    Section: 13

MULTIPLE CHOICE

41. Which of the following terms best describes the side chain of valine?

    a. acidic
    b. basic
    c. charged, polar
    d. uncharged, polar
    e. nonpolar

Answer: e Section: 1

SHORT ANSWER

42. Provide the Fischer projection of L-aspartic acid.

Answer:

$$
\begin{array}{c}
COOH \\
H_2N{-}|{-}H \\
CH_2{-}COOH
\end{array}
$$

Section: 1

43. Draw the form of L-lysine which is present at biological pH.

Answer:

$$
\begin{array}{c}
COO^- \\
H_3N^+{-}|{-}H \\
(CH_2)_4{-}NH_3{}^+
\end{array}
$$

Section: 1

44. Draw the form of L-tryptophan which is present at biological pH.

Answer:

Section: 1

45. What is the net charge of arginine in a solution of pH 1.0?

Answer: +2 Section: 1

46. Draw structures for the forms of glycine present in basic, neutral, and acidic solutions.

Answer:

$H_2N—CH_2—COO^-$ $\overset{+}{H_3N}—CH_2—COO^-$ $\overset{+}{H_3N}—CH_2—COOH$
  in basic solution   in neutral solution   in acidic solution

Section: 3

MULTIPLE CHOICE

47. Which of the following amino acids has its isoelectric point at the lowest pH?

  a. glutamic acid
  b. lysine
  c. valine
  d. glycine
  e. methionine

Answer: a Section: 4

SHORT ANSWER

48. Describe how electrophoresis works.

Answer:
A sample of an amino acid mixture is placed in the center of an acrylamide gel. Two electrodes are placed in contact with the edges of the gel and a high electrical potential is applied across the electrodes. Based on their charge, species then migrate through the gel toward the appropriate electrode and are thereby separated.

Section: 5

49. Draw the dipeptide Val-Tyr at pH 7.0.

Answer:

Section: 7

MULTIPLE CHOICE

50. When a disulfide linkage is formed, the compound containing this new linkage has been _____.

    a. hydrolyzed
    b. dehydrated
    c. electrolyzed
    d. oxidized
    e. reduced

    Answer: d Section: 7

SHORT ANSWER

51. Draw the structure of the tetrapeptide Ser-Leu-Phe-Pro at pH 7.0.

    Answer:

Section: 7

MULTIPLE CHOICE

52. What peptide remains after two sequential Edman degradations of the pentapeptide Ser-Pro-Gly-Phe-Arg?

   a. Pro-Gly-Phe
   b. Ser-Pro-Gly
   c. Gly-Phe-Arg
   d. Pro-Gly
   e. Phe-Arg

   Answer: c   Section: 12

SHORT ANSWER

53. How is the enzyme chymotrypsin used in peptide structure determination?

   Answer:
   Chymotrypsin cleaves the peptide chain at the carboxyl groups of phenylalanine, tyrosine, and tryptophan. Each of these shortened chains is then sequenced and the original peptide is sequenced by determining fragment overlap.
   Section: 12

54. Which type of protein, globular or fibrous, tends to function primarily as structural parts of an organism?

   Answer: fibrous   Section: 11

55. Give a detailed, stepwise mechanism for the formation of the activated acyl derivative from the reaction of an amino acid with N,N'-dicyclohexylcarbodiimide.

   Answer:

   Section: 9

56. Provide the structure of Thr-Gln-Met at pH7.0.

Answer:

Section: 7

57. Complete hydrolysis of a hexapeptide gives 2 Gly, Leu, Phe, Pro, and Tyr. Reaction of the peptide with phenylisothiocyanate gave the phenylthiohydantoin of Pro. Partial hydrolysis of the peptide gave the following fragments: Phe-Gly-Tyr, Gly-Phe-Gly, Pro-Leu-Gly, Leu-Gly-Phe. What is the structure of the peptide?

Answer: Pro-Leu-Gly-Phe-Gly-Tyr Section: 12

58. What is the major force responsible for the formation of an α-helix in protein secondary structure?

Answer: hydrogen bonding Section: 13

MULTIPLE CHOICE

59. What is the name of the amino acid produced when propanoic acid is subjected to the following sequence of reagents: 1. PBr_3, Br_2; 2. H_2O; 3. NH_3, Δ?

   a. alanine
   b. aspartic acid
   c. glutamic acid
   d. valine
   e. asparagine

Answer: a Section: 6

SHORT ANSWER

60. In a globular protein, would the side chain of aspartic acid most probably be oriented toward the interior of the protein or outward toward the aqueous surroundings? Explain.

Answer:
The aspartic acid side chain is negatively charged at biological pH. This negatively charged group is highly hydrophilic and will most likely be oriented so that interaction with water is possible.
Section: 14

61. In a globular protein, would the side chain of isoleucine most probably be oriented toward the interior of the protein or outward toward the aqueous surroundings? Explain.

Answer:
The isoleucine side chain is a nonpolar hydrocarbon group. This side chain is highly hydrophobic and will most likely be oriented toward the relatively nonpolar interior of the protein than toward the highly polar aqueous surroundings.
Section: 14

MULTIPLE CHOICE

62. Which of the following best describes the side chain of arginine at a pH of 11?

   a. hydrophobic
   b. nonpolar
   c. charged and polar
   d. uncharged but polar
   e. aprotic

Answer: d   Section: 3

SHORT ANSWER

63. Draw and name the amino acids which may be described as heterocyclic.

Answer:

| proline | tryptophan | histidine |

Section: 1

MULTIPLE CHOICE

64. Which of the following amino acids is <u>not</u> an aromatic compound?

   a. phenylalanine
   b. threonine
   c. histidine
   d. tryptophan
   e. tryosine

Answer: b   Section: 1

SHORT ANSWER

65. You are given a mixture that contains glutamic acid (pI = 3.2), arginine (pI = 10.8), and valine (pI = 6.0), and you subject the mixture to electrophoresis.

   a. Which amino acids migrate toward the cathode when the electrophoresis is carried out at a pH of 7.1?
   b. Which amino acids migrate toward the anode when the electrophoresis is carried out at a pH of 7.1?
   c. Which amino acid migrates farthest toward the anode at a pH of 7.1?
   d. Since amino acids themselves are colorless, how is the separation of amino acids after electrophoresis detected?

Answer:
a. arginine.
b. valine and glutamic acid.
c. glutamic acid.
d. Ninhydrin is used to locate the amino acids since the ninhydrin forms a purple derivative in the presence of an amino acid.

Section: 5

66. Rank the following amino acids in order of time of elution from a cation-exchange resin and a buffer pH of 4.0: aspartic acid, phenylalanine, arginine, and alanine.

    Answer: aspartic acid > alanine > phenylalanine > arginine   Section: 5

MULTIPLE CHOICE

67. What amino acid can be obtained by the reductive amination of ?-ketoglutaric acid, $HO_2CCH_2CH_2COCO_2H$?
    a. glycine
    b. serine
    c. lysine
    d. aspartic acid
    e. glutamic acid

    Answer: e   Section: 6

SHORT ANSWER

68. Explain the chemistry behind giving a "permanent" to a head of hair.

    Answer:
    A permanent alters the shape of hair by changing the location of the disulfide bridges within the hair. This is accomplished by first applying a reducing agent to the hair to reduce all of the existing disulfide bridges in the protein strands. Then the hair is curled into the desired shape, and an oxidizing agent is applied to form new disulfide bridges which hold the hair in the new shape.
    Section: 7

MULTIPLE CHOICE

69. In conventional peptide synthesis, the nitrogen of a given amino acid must be deactivated or blocked while the carboxyl group is activated. Which of the following reagents is used to protect the amino group of an amino acid?

    a. di-<u>tert</u>-butyl dicarbonate
    b. dicyclohexylcarbodiimide
    c. ninhydrin
    d. trifluoroacetic acid
    e. phenylisothiocyanate

    Answer: a   Section: 9

70. Consider the two peptides shown below. Which of the following methods would be able to distinguish between these two sequences?

Ala-Met-Val-Leu-Gly-Gly-Phe-Glu-Arg-Val

and

Ala-Val-Gly-Glu-Val-Met-Leu-Gly-Arg-Phe

a. One cycle of Edman's reagent.
b. Chymotrypsin digestion.
c. Complete hydrolysis of all peptide bonds.
d. Both a and b.
e. None of the above.

Answer: b   Section: 12

Chapter 22: Catalysis

MULTIPLE CHOICE

1. Which of the following is the best definition of a catalyst?

   a. It increases the rate of a chemical reaction.
   b. It decreases the rate of a reaction.
   c. It decreases activation energy.
   d. It increases activation energy.
   e. a and c.

   Answer: e Section: 98

2. Which of the following statements is not true about catalysts?

   a. They can increase the rate of a chemical reaction.
   b. They can change the mechanism of a chemical reaction.
   c. They are changed in a chemical reaction.
   d. They participate in a chemical reaction.
   e. They can decrease the activation energy of a chemical reaction.

   Answer: c Section: 98

3. Which of the following statements is/are true about catalysts?

   a. They can be used only once.
   b. They participate in the chemical reaction, but are not changed or consumed.
   c. They change the stability of reactants and products.
   d. They change the amount of product formed in a chemical reaction.
   e. All of the above.

   Answer: b Section: 98

4. Which of the following is/are true about base-catalysts?

   a. They increase the rate of a reaction by removing a proton from a reactant.
   b. They increase the rate of a reaction by providing a pathway with a more stable transition state.
   c. They facilitate the formation of a negatively charged tetrahedral intermediate.
   d. a and b only.
   e. a, b and c.

   Answer: e Section: 4

5. Which of the following is a step in the mechanism of the reaction shown below?

   a. I
   b. II
   c. III
   d. IV
   e. V

$$CH_3CH_2-Cl + \overset{\ominus}{OH} \xrightarrow{\overset{\ominus}{I}} CH_3CH_2OH + Cl^{\ominus}$$

I. $CH_3CH_2-Cl + :\overset{\ominus}{OH} \longrightarrow CH_3CH_2OH + Cl^{\ominus}$

II. $CH_3CH_2-Cl + :\overset{\ominus}{I}: \longrightarrow CH_3CH_2I + Cl^{\ominus}$

III. $\overset{\ominus}{O}-H + :\overset{\ominus}{I}: \longrightarrow HI + O^{-2}$

IV. $\underset{H}{CH_2}-CH_2-Cl + :\overset{\ominus}{OH} \longrightarrow CH_2=CH_2 + H_2O$

V. $CH_2=CH_2 + H-I \longrightarrow CH_3\overset{\oplus}{CH_2} + I^{\ominus}$

Answer: b Section: 2

6. Which of the following is an intermediate in the mechanism of hydrolysis of the ester shown below?

$$CH_3-\overset{\overset{\displaystyle O}{\|}}{C}-OCH_3 + H_2O \overset{H^+}{\rightleftharpoons} CH_3-\overset{\overset{\displaystyle O}{\|}}{C}-OH + CH_3OH$$

a.
$$CH_3-\overset{\overset{\displaystyle \oplus}{\overset{\displaystyle OH}{\|}}}{C}-OH$$

b.
$$CH_3-\overset{\overset{\displaystyle OH}{|}}{\underset{\underset{\displaystyle OH_2}{|}}{C}}-OCH_3$$

c.
$$CH_3-\overset{\overset{\displaystyle \oplus}{\overset{\displaystyle OH}{|}}}{\underset{\underset{\displaystyle OH}{|}}{C}}-OCH_3$$

d.
$$CH_3-\overset{\overset{\displaystyle \oplus}{\overset{\displaystyle OH}{\|}}}{C}-OCH_3$$

e. All of the above

Answer: e Section: 3

7. Which of the following represents the first step(s) in the mechanism involving general-base catalysis for the reaction shown below?

   a. I
   b. II
   c. III
   d. IV
   e. V

Answer: a   Section: 4

8. Which of the following is not true about metal-ion catalysts?

   a. They coordinate with atoms with nonbonded electrons.
   b. They can make a reaction center more susceptible to receiving electrons and stabilize a developing charge on the transition state.
   c. They can make a leaving group a weaker base and therefore a better leaving group.
   d. They act as Lewis bases.
   e. They can increase the electrophilicity of a reaction center.

Answer: d   Section: 5

9. Which of the following is an intermediate in the metal-ion catalyzed reaction shown below?

   a. I
   b. II
   c. III
   d. IV
   e. V

$$H_2N-CH_2\overset{O}{\overset{\|}{C}}OCH_2CH_3 \xrightarrow{Co^{+3}} CH_3CH_2OH + H_2NCH_2\overset{O}{\overset{\|}{C}}NHCH_2\overset{O}{\overset{\|}{C}}OCH_2CH_3$$

I.

II.

III.

IV.

V.

Answer: b Section: 5

10. Which of the following are affected when an intramolecular catalyst is used?

   a. Number of collisions between the two per unit time
   b. Fraction of collisions with sufficient energy
   c. Fraction of collisions with proper orientation
   d. Rate of the reaction
   e. All of the above

Answer: e Section: 6

11. What is the main purpose for using metal-ions as catalysts in the following hydrolysis reaction?

    a. Increases the rate by making water a stronger base, thereby increasing its electrophilicity
    b. Increases the rate by making water a stronger base, thereby increasing its nucleophilicity
    c. Increases the rate by making water a stronger acid, thereby increasing its nucleophilicity
    d. Increases the rate by making water a stronger acid, thereby increasing its electrophilicity
    e. Increases the rate by making water a stronger acid, thereby decreasing its nucleophilicity

Answer: c Section: 5

12. Which of the following is true of globular proteins?

    a. They may be proteins with predominantly quaternary structure.
    b. They may be circular strands of short peptides.
    c. They may be three-dimensional chains of short peptides.
    d. They may be enzymes that catalyze biological reactions.
    e. They may be large strands of proteins and amylopectins.

Answer: d Section: 8

13. How do enzymes differ from nonbiological catalysts?

    a. They have specificity for the substrate.
    b. They have lower molecular weights than nonbiological catalysts.
    c. They require higher pressure to be functional.
    d. They require lower pH to be functional.
    e. They are more polar than nonbiological catalysts.

Answer: a Section: 8

14. What does the model termed "lock-and-key" stand for?

    a. The hydrogen bonds in the α-helix fit each other like a key fits a lock.
    b. The amino acids in a peptide chain fit each other like a key fits a lock.
    c. The substrate fits the active site of the enzyme like a key fits a lock.
    d. The nonpolar sites of proteins fit each other like a key fits a lock.
    e. The N-terminal and the C-terminal ends of a protein fit each other like a key fits a lock.

    Answer: c Section: 8

15. Which of the following statements account for the remarkable catalytic ability of enzymes?

    a. Proper positioning and specificity of active site and substrate
    b. Proper positioning of amino acid side chains that serve as catalytic groups
    c. Ability of the groups on the enzyme to stabilize transition states
    d. a and b
    e. All of the above

    Answer: e Section: 8

16. Which of the following is not true about the enzyme carboxypeptidase A?

    a. Arg 145 and Tyr 248 are involved in the catalytic process of the enzyme.
    b. The metal ion in carboxypeptidase A is partially liganded to the oxygen of the carbonyl group that will be hydrolyzed.
    c. Carboxypeptidase A catalyzes the hydrolysis of the C-terminal peptide bond in proteins.
    d. Carboxypeptidase A is a metalloenzyme.
    e. The metal ion in carboxypeptidase is a Cu^{+2}.

    Answer: e Section: 9

17. Which of the labeled bonds in the following peptide chain will break when carboxypeptidase A is used to hydrolyze the chain?

$$
\begin{array}{ccccccc}
 & O & & O & & O \\
5 & \|4 & 3 & \|2 & 1 & \| & \ominus \\
-NH-CH-C- & NH-CH-C- & NH-CH-C- & O \\
| & | & | \\
R & R & R
\end{array}
$$

    a. 1
    b. 2
    c. 3
    d. 4
    e. 5

Answer: b Section: 9

18. Which of the following is not true about serine proteases?

    a. They are called serine proteases because they are able to break the peptide chain at the serine linkage.
    b. Endopeptidases and serine proteases are synonyms with each other.
    c. Trypsin, chymotrypsin, and elastase are all serine proteases.
    d. Serine proteases catalyze the hydrolysis of protein peptide chains.
    e. They are called serine proteases because they all have a serine residue that participates in the catalytic process.

Answer: a Section: 9

19. Which of the following sets of amino acids is found in all serine proteases?

    a. Arginine, histidine, serine
    b. Arginine, aspartate, serine
    c. Proline, histidine, serine
    d. Aspartate, histidine, serine
    e. Aspartate, proline, serine

Answer: d Section: 9

20. Which of the following is the main difference among the active sites of all serine proteases?

    a. The presence of serine residue at the active site
    b. The size and charge of the pocket at the active site
    c. The presence of histidine residue at the active site
    d. The presence of aspartate at the active site
    e. The hydrolysis of the peptide bond

Answer: b Section: 9

21. Which of the following intermediates is produced in the second step of the mechanism for bovine chymotrypsin hydrolysis of a peptide bond?

    a. Epoxy-enzyme intermediate
    b. Carboxy-enzyme intermediate
    c. Acyl-enzyme intermediate
    d. Phenoxy-enzyme intermediate
    e. Oxy-enzyme intermediate

    Answer: c Section: 9

22. Which of the following statements best describes what lysozyme is?

    a. An enzyme that destroys the cell walls of bacteria
    b. An enzyme that repairs the cell walls of bacteria
    c. An enzyme that hydrolyzes the NAM-NAG bond
    d. a and c
    e. b and c

    Answer: d Section: 9

23. Which of the following residues are found at the active site of lysozyme?

    a. His 57, Asp 102
    b. His 57, Asp 52
    c. Glu 35, His 57
    d. Asp 102, Glu 35
    e. Glu 35, Asp 52

    Answer: e Section: 9

24. After lysozyme binds its substrate, the enzyme undergoes a conformational change that distorts residue D in the substrate. Which of the following best describes this distortion?

    a. From a chair to a half-chair conformation
    b. From a half-chair to a chair conformation
    c. From a chair to a boat conformation
    d. From a boat to a chair conformation
    e. From a boat to a half-chair conformation

    Answer: a Section: 9

25. Which of the following is not true about the enzyme aldolase?

    a. It catalyzes the cleavage of D-glucose into two molecules each containing 3-carbons.
    b. It catalyzes the cleavage of D-galactose into two molecules each containing 3-carbons.
    c. It catalyzes the cleavage of a six carbon compound into two three carbon compounds.
    d. It catalyzes the cleavage of D-fructose 1,6-diphosphate into glyceraldehyde-3-phosphate and dihydroxyacetone phosphate.
    e. The enzyme is called aldolase because the reverse reaction is an aldol condensation.

    Answer: b   Section: 9

26. Which of the following residues acts as a general-base catalyst in the cleavage of the C_3-C_4 bond in the mechanism of the aldolase-catalyzed reaction?

    a. Methionine
    b. Histidine
    c. Cysteine
    d. Asparagine
    e. Lysine

    Answer: c   Section: 9

27. Which of the following residues acts as a general-acid catalyst in aldolase but as a general-base catalyst in chymotrypsin?

    a. Methionine
    b. Histidine
    c. Cysteine
    d. Asparagine
    e. Lysine

    Answer: b   Section: 9

28. What are transition-state analogs?

    a. Artificial ligands that are structurally similar to the transition state of the enzyme-catalyzed reaction
    b. Artificial ligands that are used as antigens to stimulate the synthesis of complementary antibodies
    c. Artificial ligands that are used as antigens to inhibit the synthesis of complementary antibodies
    d. a and b
    e. a and c

    Answer: d   Section: 10

SHORT ANSWER

29. List four common types of catalysts used in organic reactions and give one example of each.

Answer:

1. Nucleophilic catalysts: Ex $\overset{\ominus}{I}$, :N⏜:NH

2. Acid catalysts: Ex. $\overset{\oplus}{H}$, $AlCl_3$

3. Base catalysts: Ex. $\overset{\ominus}{O}H$, NH_3

4. Metal-ion catalysts: Ex Cu^{+2}, Al^{+3}, Co^{+3}, Cu^{+2} etc.

Section: 1

30. Propose a step-by-step mechanism for the general-acid catalyzed reaction shown below:

$$CH_3-\overset{O}{\overset{||}{C}}-OCH_3 + H_2O \overset{HA}{\rightleftharpoons} CH_3-\overset{O}{\overset{||}{C}}-OH + CH_3OH$$

Answer:

Section: 3

31. Show why typical metal-bound water has a considerably lower pKa than that of water.

    Answer:
    The metal ion binds to water making it easier for water to lose a proton, forming metal-bound hydroxide ion which is a stronger nucleophile than water. Typical pKa values for metal-bound water are between 5.7 - 12.7, much lower than that of 15.6 for water.

    $$metal\!-\!OH_2 \rightleftharpoons metal\!-\!\overset{\ominus}{OH} + \overset{\oplus}{H}$$

    Section: 5

32. What factor(s) would affect the rate of an intramolecular reaction?

    Answer:
    There are three main factors:
    1. Number of collisions between the two reacting groups per unit time
    2. Fraction of collisions with sufficient energy
    3. Fraction of collisions with proper orientation

    Section: 6

33. Explain why an intramolecular reaction resulting in the formation of five- or six-membered rings occurs more readily than the corresponding intermolecular reaction.

    Answer:
    An intramolecular reaction has the advantage of having the reacting groups tied together in the same molecule. Consequently, the reacting groups have a greater chance of colliding.

    Section: 6

34. Explain what is meant by anchimeric assistance and give an example.

    Answer:
    In Greek, anchimeric means "adjacent parts." This is the same thing as saying intramolecular catalyst.

    Ex.

    Section: 7

35. Explain the large difference in the relative rates of the following hydrolysis reactions:

I.

Rel. Rate = 1

II.

Aspirin

Rel Rate = 150

Answer:
This large difference in the relative rates is due to the fact that the ortho-carboxy substituent (in aspirin) is an intramolecular general-base catalyst. It increases the nucleophilicity of H_2O, thereby increasing the rate of formation of the tetrahedral intermediate.

Section: 7

36. Explain the phenomenon known as molecular recognition. Give an example.

Answer:
Molecular recognition is the recognition of one molecule by another molecule.

Ex.
The specificity of the active site of the enzyme for its substrate. This results from the conformation of the protein and the particular amino acid side chains that make up the active site. There are many other kinds of molecular recognition other than enzyme-substrate; hormone-receptors and inhibitor-enzyme.

Section: 8

37. Explain what is meant by the induced fit model. Give an example.

Answer:
Binding energy from the substrate-enzyme interaction is used to induce a conformational change in the enzyme to better fit the substrate.

Ex.
The enzyme carboxypeptidase A fits this model.

Section: 8

38. Explain what is meant by site-specific mutagenesis.

    Answer:
    It is a technique for determining the relationship between protein function and structure. This is done by substituting one amino acid of a protein for another and seeing what the effect is.
    Section: 9

39. What are NAM and NAG and what do they do?

    Answer:
    NAM is N-acetylmuramic acid and NAG is N-acetylglucosamine. They are molecules that make up the cell walls of bacteria.
    Section: 9

40. Why is histidine a versatile catalytic group?

    Answer:
    Because the imidazole ring in histidine has a pKa of 6.8 which is close to the physiological pH of 7.2; thus it can exist in both its acidic and basic forms.
    Section: 9

MULTIPLE CHOICE

41. "Being in the right place... can be the most powerful factor of all in determining how fast a reaction goes." In terms of reaction rate theory, this statement refers to the _____ for a reaction.

    a. probability factor
    b. energy factor
    c. resonance effect
    d. inductive effect

    Answer: a   Section: 6

42. The compound $CH_3OCH_2CH_2CH_2CH_2Br$ undergoes a solvolysis reaction is aqueous ethanol solution. Which of the following is evidence that a neighboring group effect is occuring?

    a. $CH_3CH_2CH_2CH_2Br$ reacts more slowly than $CH_3OCH_2CH_2CH_2CH_2Br$ does.
    b. The reaction is faster when the concentration of water is higher.
    c. The reaction is first order.
    d. HBr is a product of the reaction.

Answer: a   Section: 6

43. A pure sample of the <u>trans</u>-2-bromocyclopentanol enantiomer shown below reacts with HCl. The reaction proceeds by way of a bromonium ion. What is/are the product(s) of the reaction?

    a. I only
    b. II only
    c. I and III
    d. II and IV

Answer: c   Section: 7

44. Which of the following sets of parameters are <u>all</u> altered when a reaction is carried out in the presence of a catalyst?

    a. $\Delta G°$, $\Delta H°$, K_{eq}
    b. E_a, free energy of activation, k_{rate}
    c. $\Delta G°$, k_{rate}, K_{eq}
    d. $\Delta H°$, $DS°$, E_a
    e. $\Delta H°$, $\Delta S°$, k_{rate}

Answer: b   Section: 98

SHORT ANSWER

45. Offer an explanation of how iodide catalyzes the hydrolysis of 1-chloropropane in aqueous base.

Answer:
The iodide acts as a nucloephile and displaces the chloride in an S_N2 process. Since iodide is a better nucleophile than hydroxide in aqueous solution, this reaction proceeds faster than the uncatalyzed reaction. In the second step of the catalyzed reaction, hydroxide displaces iodide in another S_N2 process. This displacement occurs faster than the displacement of chloride by hydroxide in the uncatalyzed reaction since iodide is a better leaving group. Overall the two-step catalyzed process occurs more rapidly than the single-step uncatalyzed displacement.

Section: 2

46. Hydrolysis of esters occurs more rapidly in the presence of catalytic imidazole. Explain this phenomenon and provide the structure of the acyl imidazole intermediate which is formed in the catalyzed process.

Answer:
Imidazole is a better nucleophile than water, so imidazole reacts faster than water with the ester carbonyl. The acyl imidazole that eventually results from this attack is particularly susceptible to attack by water because the positively charged nitrogen makes imidazole a very good leaving group.

Section: 2

MULTIPLE CHOICE

47. Which of the following statements about how acid catalyzes the hydrolysis of esters is (are) correct?

    a. The acid catalyst increases the rate of formation of the tetrahedral intermediate by protonating the carbonyl and thereby increasing the reactivity of the carbonyl group.
    b. The acid catalyst increases the rate of the reaction by changing the basicity of the group eliminated when the tetrahedral intermediate collapses.
    c. The acid catalyst protonates the water, the nucleophile in the reaction, and thereby makes it a stronger nucleophile.
    d. Both a and b.
    e. All of the above.

    Answer: d   Section: 3

48. Which of the following statements concerning general-acid catalysis is (are) correct?

    a. The proton is transferred to the reactant during the slow step of the reaction.
    b. The transition state of the slow step of the reaction has a partially transferred proton instead of a fully transferred proton.
    c. Typically, a general-acid catalyst is stronger than a specific-acid catalyst.
    d. Both a and b.
    e. All of the above.

    Answer: d   Section: 3

SHORT ANSWER

49. The reaction below occurs by a general-acid catalyzed mechanism. Use curved arrows to illustrate the ring-forming step of the mechanism.

Answer:

Section: 3

50. Rank the following compounds in increasing order of reactivity in the intramolecular displacement of p-bromophenolate to form a cyclic anhydride. Explain your ranking.

Answer:
B < C < A. B is slower than C since it has more conformations in which the reacting groups are too separated for reaction to occur. A is the fastest since the reacting groups are held proximate in space by the rigid double bond.

Section: 6

51. Rank the following alkyl chlorides in increasing order of reactivity in aqueous ethanol, and explain why one of the compounds is 70,000 times more reactive than the other two.

Answer:

B < A < C. C is the most reactive since the nucleophilic sulfur atom in **C** is set up to displace the chloride ion by an intramolecular backside attack. The resulting sulfonium ion is then attacked by water or ethanol.

Section: 6

MULTIPLE CHOICE

52. The enzyme lysozyme exhibits greatest activity in solutions where the pH is maintained between 3.8 and 6.7. Which of the following conclusions about the Asp 52 and Glu 35 residues in the enzyme's active site can be reached?

   a. Lysozyme is maximally active when both Asp 52 and Glu 35 are in their acidic forms.
   b. Lysozyme is maximally active when both Asp 52 and Glu 35 are in their basic forms.
   c. Lysozyme is maximally active when both Asp 52 and Glu 35 are in their neutral forms.
   d. Lysozyme is maximally active when Asp 52 is in its acidic form and Glu 35 is in its basic form.
   e. Lysozyme is maximally active when Asp 52 is in its basic form and Glu 35 is in its acidic form.

Answer: e   Section: 9

Chapter 23:   The Organic Mechanisms of the Coenzymes

MULTIPLE CHOICE

1. Which of the following statements best describes what cofactors are?

    a. They are metal ions that assist enzymes in carrying out their catalytic reactions.
    b. They are organic molecules that assist enzymes in carrying out their catalytic reactions.
    c. They are organic molecules called coenzymes.
    d. They are organic molecules derived from compounds commonly known as vitamins.
    e. All of the above.

    Answer: e Section: 98

2. Which of the following statements is <u>not</u> true of coenzymes?

    a. Coenzymes catalyze chemical reactions similar to enzymes.
    b. Coenzymes may function as oxidizing and reducing agents.
    c. Coenzymes play a variety of chemical roles exactly like those played by the amino acid side chains of enzymes.
    d. Coenzymes may function as nucleophiles or strong bases needed in a reaction.
    e. Coenzymes may function as activators needed in a reaction.

    Answer: c Section: 98

3. Which of the following is/are coenzyme(s) formed from the vitamin niacin?

    a. NAD^+
    b. $NADP^+$
    c. NADH
    d. NADPH
    e. All of the above

    Answer: e Section: 2

4. Which of the following is/are the coenzyme(s) formed from the vitamin thiamine (B_1)?

    a. TPP
    b. CoASH
    c. PLP
    d. THF
    e. FAD

    Answer: a Section: 4

5. Which of the following is/are water-soluble vitamins?

   a. Vitamin B_{12}
   b. Riboflavin (vitamin B_2)
   c. Thiamine (vitamin B_1)
   d. Pyridoxine (vitamin B_6)
   e. All of the above

   Answer: e Section: 98

6. Which of the following is <u>not</u> a lipid-soluble vitamin?

   a. Vitamin A
   b. Vitamin D
   c. Vitamin E
   d. Vitamin C
   e. Vitamin K

   Answer: d Section: 98

7. Which of the following vitamins is the only lipid-soluble vitamin that is known to function as a coenzyme?

   a. Vitamin A
   b. Vitamin D
   c. Vitamin B_1
   d. Vitamin C
   e. Vitamin K

   Answer: e Section: 98

8. Which of the following vitamins is the only water-soluble vitamin that <u>does</u> <u>not</u> function as a coenzyme?

   a. Vitamin B_{12}
   b. Riboflavin (B_2)
   c. Vitamin C
   d. Thiamine (B_1)
   e. Pyridoxine (B_6)

   Answer: c Section: 98

9. Which of the following vitamins are radical inhibitors or antioxidants?

   a. Niacin
   b. Vitamins B_1 and B_{12}
   c. Vitamins B_2 and B_6
   d. Vitamins A and D
   e. Vitamins C and E

   Answer: e Section: 98

10. Which of the following vitamins can be synthesized by all mammals except humans and guinea pigs?

    a. Vitamin B_{12}
    b. Riboflavin (B_2)
    c. Vitamin C
    d. Thiamine (B_1)
    e. Pyridoxine (B_6)

    Answer: c Section: 98

11. In which of the following classes of reactions is a coenzyme required?

    a. Oxidation reactions
    b. Two carbon transfer
    c. Acyl transfer
    d. Reduction reactions
    e. a, b, and d

    Answer: e Section: 1

12. Which of the following statements is/are true about metabolism?

    a. Reactions that living organisms carry out to obtain energy and to synthesize required compounds
    b. Reactions that organic chemists carry out to obtain energy and to synthesize required compounds
    c. Reactions that consist of catabolic and anabolic reactions
    d. a and c
    e. b and c

    Answer: d Section: 1

13. Which of the following is not true about catabolic reactions?

    a. Catabolic reactions require NAD^+ for a coenzyme.
    b. Catabolic reactions are most often oxidation reactions.
    c. Catabolic reactions are most often reduction reactions.
    d. Catabolic reactions require an oxidizing coenzyme.
    e. Catabolic reactions break down complex molecules into simpler ones.

    Answer: c Section: 1

14. At what position of the pyridine ring in the following coenzyme will oxidation-reduction reactions take place?

   a. 1
   b. 2
   c. 3
   d. 4
   e. 5

Answer: d Section: 2

15. What is the <u>major</u> organic product of this oxidation-reduction reaction?

a. R—C=O
     |
     H

b. CH_3—C—OH_S
       ||
       O

c. CH_3—C—OH_R
       ||
       O

d. CH_3—C=O
       |
       H_R

e. CH_3—C=O
       |
       H_S

Answer: e Section: 2

16. Generally speaking, which of the following oxidation-reduction reactions utilize NAD^+ and $NADP^+$ as oxidation coenzymes?

    a. Alcohols oxidized to ketones
    b. Alcohols oxidized to aldehydes
    c. Alcohols oxidized to carboxylic acids
    d. a and b
    e. a, b, and c

    Answer: e Section: 2

17. What are the products obtained from the reaction of NADH with FMN catalyzed by NADH dehydrogenase as shown below?

    $$NADH + FMN \xrightarrow[\text{dehydrogenase}]{\text{NADH}} ? + ?$$

    a. $NAD^+ + FMNH_2$
    b. $NAD^+ + FMNH$
    c. $NAF + DHN$
    d. $NAF + DNH_2$

    e. $N_2A_2D_2 + FMNH_2$

    Answer: a Section: 3

18. At what position on the following coenzyme (FAD/FMN) will oxidation-reduction reactions take place?

    a. 6
    b. 2
    c. 3
    d. 4
    e. 5

    Answer: e Section: 3

19. Which of the following substrates requires the coenzyme thiamine pyrophosphate (TPP) for catalyzing its decarboxylation reaction?

    a.

β-keto acid, $CH_3-\overset{\displaystyle O}{\overset{\|}{C}}-CH_2-COO^{\ominus}$

    b.

α-keto acid, $CH_3-\overset{\displaystyle O}{\overset{\|}{C}}-COO^{\ominus}$

    c.

f-keto acid, $CH_3-\overset{\displaystyle O}{\overset{\|}{C}}-CH_2CH_2-COO^{\ominus}$

    d.

α-hydroxy acid, $CH_3-\overset{\displaystyle OH}{\overset{|}{CH}}-COO^{\ominus}$

    e.

β-hydroxy acid, $CH_3-\overset{\displaystyle OH}{\overset{|}{CH}}-CH_2-COO^{\ominus}$

Answer: b   Section: 4

20. Which of the following statements best describes what an "electron sink" is?

    a. A free radical initiator for enzyme catalysis
    b. A site for electron localization
    c. A site for electron delocalization
    d. A site that donates electrons for enzyme catalysis
    e. A site that accepts electrons for enzyme catalysis

Answer: c   Section: 4

21. In addition to TPP, which of the following coenzymes is/are required to facilitate the following reaction?

$$CH_3-\overset{\displaystyle O}{\overset{\|}{C}}-COO^- + CoASH \xrightarrow[\text{system}]{\text{Pyruvate dehydrogenase}} CH_3-\overset{\displaystyle O}{\overset{\|}{C}}-S-CoA + CO_2$$

Pyruvate   Acetyl CoA

    a. Lipoate
    b. Coenzyme A
    c. FAD
    d. NAD^+
    e. All of the above

Answer: e   Section: 4

22. Which of the following coenzymes is required for the enzyme catalyzed reaction shown below?

$$\text{Xylose-5-P + Ribose-5-P} \xrightarrow{\text{transketolase}} \text{Glyceraldehyde-3-P + Sedoheptulose-7-P}$$

    a. TPP
    b. PLC
    c. THF
    d. NADH
    e. NADPH

Answer: a Section: 4

23. Which of the following reactions does the coenzyme biotin catalyze?

    a. Decarboxylation of an α-keto acid
    b. Decarboxylation of an α-carboxy acid
    c. Carboxylation of an α-carboxy acid
    d. Decarboxylation of a carbon adjacent to an ester or keto group
    e. Carboxylation of a carbon adjacent to an ester or keto group

Answer: e Section: 5

24. Which of the following are required species for biotin-requiring enzymes?

    a. ATP
    b. Mg^{+2}
    c. HCO_3^-
    d. a and b
    e. a, b and c

Answer: e Section: 5

25. Which of the following is the sequence of steps followed by biotin-requiring enzymes, as listed below?
    I.   Transfer of the carboxy group from carboxybiotin to the substrate
    II.   Activation of bicarbonate by ATP
    III. Formation of carboxybiotin

    a. I, II, III
    b. I, III, II
    c. II, I, III
    d. II, III, I
    e. III, II, I

Answer: d Section: 5

26. What does the term "Pyridoxal" stand for?

    a. Pyridine aldehyde
    b. Pyrrol oxyaldehyde
    c. Pyrrolidine oxyaldehyde
    d. Peroxy aldehyde
    e. Pyrimidoxy aldehyde

Answer: a   Section: 6

27. Which of the following reactions of amino acids is <u>not</u> catalyzed by PLP-requiring enzymes?

    a. decarboxylation
    b. transamination
    c. α,β-elimination
    d. α-keto esterification
    e. $C\alpha$-$C\beta$ bond cleavage

Answer: d   Section: 6

28. Which of the following reactions is called an α,β-elimination of an amino acid?

    a. I
    b. II
    c. III
    d. IV
    e. V

I. $CH_3-CH_2-COO^{\ominus}$ (with $\overset{\oplus}{N}H_3$) $\xrightarrow[PLP]{enzyme}$ $CH_3CH_2\overset{\oplus}{N}H_3 + CO_2$

II. $CH_2-CH-COO^{\ominus}$ (with X, $\overset{\oplus}{N}H_3$) $\xrightarrow[PLP]{enzyme}$ $CH_3\overset{O}{\overset{\|}{C}}COO^{\ominus} + X^{\ominus} + \overset{\oplus}{N}H_4$

III. $CH_2-COO^{\ominus} + {}^{\ominus}OOC-CH_2\overset{O}{\overset{\|}{C}}-COO^{\ominus}$ (with $\overset{\oplus}{N}H_3$) $\xrightarrow[PLP]{enzyme}$ $CH_3-\overset{O}{\overset{\|}{C}}-COO^{\ominus} + {}^{\ominus}OOC-CH_2-\overset{\overset{\oplus}{N}H_3}{\underset{}{C}H}-COO^{\ominus}$

IV. $R-\overset{H}{\underset{\overset{\oplus}{N}H_3}{C}}-\overset{O}{\overset{\|}{C}}-O^{\ominus}$ $\xrightarrow[PLP]{enzyme}$ $R-\overset{H}{\underset{\overset{\oplus}{N}H_3}{C}}-COO^{\ominus} + {}^{\ominus}OOC-\overset{H}{\underset{\overset{\oplus}{N}H_3}{C}}-R$

V. $HO-CH-CH-COO^{\ominus}$ (with R, $\overset{\oplus}{N}H_3$) $\xrightarrow[PLP]{enzyme}$ $HC{=}O$ (with R) $+ CH_2COO^{\ominus}$ (with $\overset{\oplus}{N}H_3$)

Answer: b   Section: 6

29. Which of the following statements is <u>not</u> true about the structure of vitamin B_{12}?

    a. The structure was determined by Dorothy Hodgkin in 1955 using X-ray crystallography.
    b. The structure has an adenosyl group.
    c. The structure has a cyano group, HO^-, or H_2O.
    d. The cyano group is coordinated to cobalt.
    e. The structure cannot be synthesized by plants and animals.

    Answer: b Section: 7

30. Which of the following best represents the chemical reaction catalyzed by B_{12}-dependent enzymes?

    Answer: a Section: 7

31. Which of the following statements is <u>not</u> true about the coenzyme tetrahydrofolate (THF)?

    a. There are six different THF-coenzymes.
    b. The coenzyme is used by enzyme-catalyzing reactions that donate a group containing a single carbon to their substrate.
    c. THF results from the reduction of two double bonds of the precursor vitamin (folate).
    d. Mammals synthesize the vitamin (folate) but bacteria cannot.
    e. Homocysteine methyl transferase and glycinamide ribonucleotide are examples of enzymes that require THF-coenzyme.

    Answer: d Section: 8

32. The thymine molecule used for the biosynthesis of DNA is synthesized from uracil as shown below.

    The structure of dTMP is a methylated dUMP. What is the structure of dTMP?

    a. I
    b. II
    c. III
    d. IV
    e. V

    Answer: e Section: 8

33. Which of the following statements is/are true about 5-fluorouracil,

    a. It is an anticancer drug.
    b. It inhibits the enzyme thymidylate synthase.
    c. It is an example of a mechanism-based inhibitor.
    d. It is an example of a suicide inhibitor.
    e. All of the above.

Answer: e Section: 8

34. Which of the following statements is <u>not</u> true about vitamin K?

    a. The letter K comes from koagulation.
    b. Vitamin K is required for proper Ca^{+2} binding in blood clotting.
    c. Kitase is the active coenzyme of vitamin K.
    d. Vitamin K is synthesized by intestinal bacteria.
    e. Vitamin K is formed in leaves of green plants.

Answer: c Section: 9

35. Which of the following statements is/are true about the vitamin KH_2?

    a. It is the active coenzyme form of vitamin K.
    b. Vitamin K is a quinone while vitamin KH_2 is a hydroquinone.
    c. Vitamin KH_2 is the coenzyme for the enzyme that catalyzes the carboxylation of glutamate side chains in proteins to form Å -carboxyglutamates.
    d. a and c.
    e. a, b, and c.

Answer: e Section: 9

SHORT ANSWER

36. What are holoenzymes and apoenzymes?

    Answer:
    Holoenzymes are enzymes with their cofactor while apoenzymes are enzymes without their cofactor.

    Section: 98

37. Which type of vitamins, water-soluble or lipid-soluble, can we overdose on?

   Answer:
   One can overdose on lipid-soluble vitamins because they are not easily eliminated by the body and can accumulate in cell membranes and other nonpolar components of the body.

   Section: 98

38. Explain what is meant by a nucleotide and give an example.

   Answer:
   A nucleotide is composed of a heterocyclic molecule attached to the β-position of a phosphorylated ribose.
   Ex.
   NAD^+ (nicotinamide adenine dinucleotide) is composed of two nucleotides linked together through their phosphate group.

   Section: 2

39. What are catabolic and anabolic reactions?

   Answer:
   Catabolic reactions are reactions where complex molecules break down into simple molecules and energy.

   Anabolic reactions are reactions where simple molecules react with energy to produce complex biomolecules.

   Section: 1

40. Can an NAD^+/NADH dependent enzyme tell the difference between the pro-R hydrogen and pro-S hydrogen of a substrate? Explain.

   Answer:
   The enzyme can tell the difference between R and S hydrogens because the enzyme has a specific binding site for its coenzyme which blocks one of the coenzyme's sides, leaving the other side open for the substrate to bind to. For example, alcohol dehydrogenase removes only the pro-R hydrogen of ethanol.

   Section: 2

41. Explain what is meant by a prosthetic group. Give an example.

Answer:
A tightly bound coenzyme is known as a prosthetic group.
Ex.
Flavin adenine dinucleotide (FAD) or
Flavin mononucleotide (FMN)
Section: 3

42. Propose a mechanism for the following reaction.

Answer:

Section: 3

43. Propose a mechanism for the following reaction.

Answer:

Section: 4

44. Show the transimination (trans-Schiffization) reaction that happens in all PLP-requiring enzymes in which the amino acid substrate forms a Schiff base with PLP, thereby freeing the lysine side chain.

Answer:

Section: 6

45. Propose a mechanism for the following coenzyme B_{12}-dependent enzyme-catalyzed reaction.

Answer:

Section: 7

46. Propose a mechanism for the vitamin KH_2-catalyzed carboxylation of glutamate shown below:

Answer:

Section: 9

MULTIPLE CHOICE

47. Which of the following vitamins catalyzes acyl transfer reactions and requires CoASH as a coenzyme?

   a. niacin
   b. thiamine
   c. vitamin B_{12}
   d. lipoate
   e. panthothenate

Answer: e   Section: 98

48. Which of the following vitamins catalyzes one-carbon transfer reactions and requires THF as a coenzyme?
    a. biotin
    b. thiamine
    c. folic acid
    d. ascorbic acid
    e. panthothenate

Answer: c   Section: 98

SHORT ANSWER

49. List the four stages of catabolism and briefly explain the significance of each.

Answer:
The first stage of catabolism is called digestion. In this stage, fats, carbohydrates, and proteins are hydrolyzed into fatty acids, monosaccharides, and amino acids respectively. The second stage is one in which these hydrolytic products are converted into compounds which can enter the citric acid cycle (acetyl-CoA, pyruvate, citric acis cycle intermediates). The third stage of catabolism is the citric acid cycle itself. For every acetyl-CoA that enters the cycle, two molecules of carbon dioxide are formed. The final stage of catabolism is called oxidative phosphorylation. Herein, every NADH that was formed in the earlier stages is converted into three ATPs.
Section: 1

MULTIPLE CHOICE

50. In a process known as oxidative phosphorylation, every NADH that is formed in carrying out oxidation reactions in the early stages of catabolism is converted into _____.

    a. three ATPs
    b. two ATPs
    c. four FADHs
    d. six acetyl-CoAs
    e. none of the above

Answer: a   Section: 1

51. Which of the following statements regarding glycolysis is correct?

    a. Glycolysis is another name for the Krebs cycle.
    b. Disaccharides can enter glycolysis without being hydrolyzed.
    c. Glycolysis occurs primarily in the fluid outside the cell.
    d. Glycolysis is the series of reactions which produces 2 moles of pyruvate from each mole of D-glucose.
    e. Glycolysis is the series of reactions which is responsible for the complete oxidation of pyruvate and acetyl-CoA to carbon dioxide.

    Answer: d   Section: 2

SHORT ANSWER

52. What structural feature of enzymes that catalyze redox reactions requires the presence of a coenzyme for reaction to occur?

    Answer:
    An enzyme that catalyzes a redox reaction requires a coenzyme because none of the amino acid side chains can act as oxidizing or reducing agents.
    Section: 2

53. What are the main structural and functional differences between NAD^+/NADH and $NADP^+$/NADPH?

    Answer:
    The only way in which $NADP^+$ differs structurally from NAD^+ is the phosphate group bonded to the 2'-OH group of the ribose of the adenine nucleotide. Functionally, NAD^+ and NADH are generally used as coenzymes in catabolic reactions while $NADP^+$ and NADPH are generally used as coenzymes in anabolic reactions.
    Section: 2

54. Explain why a cell typically maintains its [NAD^+]/[NADH] ratio near 1000 while maintaining its [$NADP^+$]/[NADPH] ratio at about 0.01.

    Answer:
    The relative concentrations of coenzymes in the cell encourage binding of the appropriate coenzyme. Since NAD^+ and NADH are catabolic coenzymes, the NAD^+ concentration in the cell needs to be much greater than the NADH concentration because catabolic reactions are most often oxidation reactions. The opposite is true of $NADP^+$ and NADPH. These are anabolic coenzymes that are predominantly reduction reactions which require a higher concentration of NADPH.
    Section: 2

MULTIPLE CHOICE

55. How does NAD$^+$ serve as a catalyst in oxidation reactions?

    a. It accepts a hydride ion at the 4-position of its pyridine ring.
    b. It accepts a hydride ion at the amide carbonyl group at the 3-position.
    c. It accepts a proton at the 4-position of its pyridine ring.
    d. It accepts a proton at the amide carbonyl group at the 3-position.
    e. none of the above.

    Answer: a   Section: 2

SHORT ANSWER

56. Through what enzymatic pathway is thymine biosynthesized?

    Answer:
    The Ts used for the biosynthesis of DNA are synthesized from Us (uracils) by thymidylate synthase and the coenzyme N^5,N^{10}-methylene-THF. Even though the only structural difference between a U and a T is a methyl group, a T is synthesized by first transferring a methylene group to a U.

    Section: 8

Chapter 24: Lipids

MULTIPLE CHOICE

1. Which of the following statements is <u>not</u> true about fatty acids?

   a. Fatty acids are carboxylic acids with long hydrocarbon side chains.
   b. The double bonds in unsaturated fatty acids are always conjugated.
   c. Most naturally occuring fatty acids contain even numbers of carbons and are unbranched.
   d. Fatty acids can be saturated or unsaturated.
   e. Physical properties of fatty acids depend on the length of the hydrocarbon chain and the degree of unsaturation.

   Answer: b Section: 1

2. Which of the following statements is <u>true</u> for both cholesterol and vitamin A?

   a. Both are vitamins.
   b. Both are proteins.
   c. Both are lipids.
   d. Both are steroids.
   e. Both are fatty acids.

   Answer: c Section: 1

3. Which of the following best explains why the melting points of saturated fats increase with increasing molecular weight?

   a. Decreased polarity
   b. Decreased hydrogen bonding
   c. Increased hydrogen bonding
   d. Decreased intermolecular van der Waal's interactions
   e. Increased intermolecular van der Waal's interactions

   Answer: e Section: 1

4. Which of the following statements best describes the structure of waxes?

   a. Long-chain unsaturated carboxylic acids.
   b. Long-chain saturated carboxylic acids.
   c. Long-chain esters.
   d. Short-chain esters.
   e. Long-chain anhydrides.

   Answer: c Section: 2

5. Which of the following molecules is known as a triglyceride?

a.

$$
\begin{array}{l}
CH_2-O-\overset{\overset{\displaystyle O}{\|}}{C}-R\\
\;\;\;|\qquad\;\;\overset{\displaystyle O}{\|}\\
CH-O-\overset{\|}{C}-R\\
\;\;\;|\qquad\qquad\overset{\displaystyle O}{\|}\\
CH_2-O-\overset{\|}{C}-R
\end{array}
$$

b.

$$
\begin{array}{l}
\qquad\qquad\overset{\displaystyle O}{\|}\\
CH_2-\overset{\|}{C}-OR\\
|\qquad\quad\overset{\displaystyle O}{\|}\\
CH-\overset{\|}{C}-OR\\
|\qquad\qquad\overset{\displaystyle O}{\|}\\
CH_2-\overset{\|}{C}-OR
\end{array}
$$

c.

$$
\begin{array}{l}
\qquad\quad\overset{\displaystyle O}{\|}\\
CH_2-\overset{\|}{C}-R\\
|\qquad\;\;\overset{\displaystyle O}{\|}\\
CH-\overset{\|}{C}-R\\
|\qquad\quad\overset{\displaystyle O}{\|}\\
CH_2-\overset{\|}{C}-R
\end{array}
$$

d.

$$
\begin{array}{l}
\qquad\qquad\overset{\displaystyle O}{\|}\\
CH_2-O-\overset{\|}{C}-OH\\
|\qquad\qquad\overset{\displaystyle O}{\|}\\
CH-O-\overset{\|}{C}-OH\\
|\qquad\qquad\;\overset{\displaystyle O}{\|}\\
CH_2-O-\overset{\|}{C}-OH
\end{array}
$$

e.

$$
\begin{array}{c}
\text{CH}_2 \cdot \text{OO}-\overset{\displaystyle\overset{O}{\|}}{\text{C}}-\text{R} \\
| \\
\text{CH}-\text{OO}-\overset{\displaystyle\overset{O}{\|}}{\text{C}}-\text{R} \\
| \\
\text{CH}_2 \cdot \text{OO}-\overset{\displaystyle\overset{O}{\|}}{\text{C}}-\text{R}
\end{array}
$$

Answer: a   Section: 3

6. Which of the following statements is <u>not</u> true about triacylglycerols?

   a. When solids and semisolids at room temperature, they are called fats.
   b. When liquids at room temperature, they are called oils.
   c. When hydrolyzed, they produce glycerol and carboxylate salts.
   d. Triacylglycerols with low melting points are composed of saturated fatty acids, causing them to be liquids at room temperature.
   e. Commonly known soap is the salt of a long-chain fatty acid which can be obtained from triacylglycerols.

   Answer: d   Section: 3

7. Which of the following statements is <u>not</u> true about phospholipids?

   a. They are similar to triacylglycerols except that the middle OH group of glycerol reacts with a phosphate rather than with a fatty acid.
   b. They constitute a major component of cell membranes.
   c. Phosphatidic acid is a phospholipid.
   d. The C-2 carbon of glycerol in phosphoacylglycerols has the R configuration.
   e. Cephalin and lecithin are examples of phospholipids.

   Answer: a   Section: 4

8. Which of the following molecules is known as phosphatidic acid?

a.

$$
\begin{array}{c}
\text{CH}_2-\text{O}-\overset{\overset{\displaystyle O}{\|}}{\text{C}}-R_1 \\
\overset{\overset{\displaystyle O}{\|}}{\text{CH}-\text{O}-\text{P}-\text{OH}} \\
\overset{\displaystyle |\ominus}{\underset{\displaystyle O}{}} \\
\overset{\overset{\displaystyle O}{\|}}{\text{CH}_2-\text{O}-\text{C}-R_2}
\end{array}
$$

b.

$$
\begin{array}{c}
\text{CH}_2-\text{O}-\overset{\overset{\displaystyle O}{\|}}{\text{C}}-R_1 \\
\overset{\overset{\displaystyle O}{\|}}{\text{CH}-\text{O}-\text{P}-\text{OH}} \\
\overset{\displaystyle |\ominus}{\underset{\displaystyle O}{}} \\
\overset{\overset{\displaystyle O}{\|}}{\text{CH}_2-\text{O}-\text{C}-R_2}
\end{array}
$$

c.

$$
\begin{array}{c}
\text{CH}_2-\text{O}-\overset{\overset{\displaystyle O}{\|}}{\text{C}}-R_1 \\
\overset{\overset{\displaystyle O}{\|}}{\text{CH}-\text{O}-\text{C}-R_2} \\
\overset{\overset{\displaystyle O}{\|}}{\text{CH}_2-\text{O}-\text{P}-\text{OH}} \\
\overset{\displaystyle |\ominus}{\underset{\displaystyle O}{}}
\end{array}
$$

d.

$$
\begin{array}{c}
\overset{\overset{\displaystyle O}{\|}}{\text{CH}_2-\text{O}-\text{P}-\text{OH}} \\
\overset{\displaystyle |\ominus}{\underset{\displaystyle O}{}} \\
\overset{\overset{\displaystyle O}{\|}}{\text{CH}-\text{O}-\text{C}-R_1} \\
\overset{\overset{\displaystyle O}{\|}}{\text{CH}_2-\text{O}-\text{C}-R_2}
\end{array}
$$

568

e.

$$CH_2 - O - \overset{\overset{\displaystyle O}{\|}}{P} - OR$$

$$\overset{\displaystyle |\ominus}{O}$$

$$\overset{\displaystyle O}{O}$$

$$CH - O - \overset{\overset{\displaystyle O}{\|}}{C} - R_1$$

$$\overset{\displaystyle O}{O}$$

$$CH_2 - O - \overset{\overset{\displaystyle O}{\|}}{C} - R_2$$

Answer: c Section: 4

9. Which of the following is the skeleton for prostaglandins?

a. I
b. II
c. III
d. IV
e. V

I.

II.

III.

IV.

V.

Answer: c Section: 5

10. Which of the following fatty acids cannot be synthesized by mammals but must be included in the diet because it is needed to synthesize arachidonic acid which in turn will synthesize prostaglandins?

    a. I
    b. II
    c. III
    d. IV
    e. V

    I. Lauric acid   ~~~~~~~COOH

    II. Palmitic acid   ~~~~~~~~~COOH

    III. Arachidic acid   ~~~~~~~~~~COOH

    IV. Linoleic acid   ~~~~~~COOH

    V. Oleic acid   ~~~~~~~COOH

    Answer: d Section: 1

11. Why is aspirin used as an anti-inflammatory drug?

    a. Aspirin enhances the production of prostaglandins.
    b. Aspirin inhibits the production of prostaglandins.
    c. Aspirin promotes the activity of cyclooxygenase.
    d. Aspirin transfers a hydroxy group to a serine hydroxyl group of the enzyme.
    e. Aspirin causes an increase in the rate of blood clotting.

    Answer: b Section: 5

12. Which of the following statements is true about terpenes?

    a. They are a class of lipids that can be isolated from plants.
    b. They contain carbon atoms in multiples of five.
    c. They are composed of isoprene units joined together in a head-to-tail fashion.
    d. Tetraterpenes have 40 carbon atoms.
    e. All of the above.

    Answer: e Section: 6

13. How many isoprene units are in a sesquiterpene?

    a. 1
    b. 2
    c. 3
    d. 4
    e. 5

Answer: c Section: 6

14. How many isoprene units are in β-carotene?

    a. 4
    b. 5
    c. 6
    d. 7
    e. 8

Answer: e Section: 6

15. The five-carbon compound used for the biosynthesis of terpenes is called isopentenyl pyrophosphate. Which of the following is the structure of isopentenyl pyrophosphate?

a. I
b. II
c. III
d. IV
e. V

I.

II.

III.

IV.

V.

Answer: a   Section: 8

16. Squalene is an important precursor of cholesterol which is the precursor of all other steroids. Squalene is prepared from two molecules of farnesyl pyrophosphate that are joined head-to-head.

    What is the structure of farnesyl pyrophosphate?

    a. I
    b. II
    c. III
    d. IV
    e. V

Answer: c Section: 10

17. Which of the following statements is **not** true about steroids?

    a. They are lipids.
    b. They are hormones.
    c. They are synthesized in the glands and delivered by the bloodstream to tissues in order to stimulate or inhibit certain processes.
    d. All steroids contain a tetracyclic ring system.
    e. In steroids, the B, C, and D rings are all cis fused.

    Answer: e Section: 9

18. Which of the following is the precursor for <u>all</u> steroids?

    a. Cholesterol
    b. Cortisone
    c. Progesterone
    d. Estrone
    e. Androsterone

Answer: a Section: 9

19. How many chirality centers does cholesterol have?

    a. 4
    b. 5
    c. 6
    d. 7
    e. 8

Cholesterol

Answer: e Section: 9

20. Which of the following classes of hormones are involved in preparing the lining of the uterus for implantation of an ovum and are essential for the maintenance of pregnancy?

    a. Estrogen
    b. Androgen
    c. Mineralocorticoids
    d. Glucocorticoids
    e. Progestins

Answer: e Section: 9

21. To which class of hormones does the steroid cortisone belong?

    a. Estrogen
    b. Androgen
    c. Mineralocorticoids
    d. Glucocorticoids
    e. Progestins

Answer: d Section: 9

22. Which of the following statements best describes the structural difference between testosterone, a male sex hormone, and progesterone, a female hormone?

    a. They only differ in the substituent at C-3.
    b. They only differ in the substituent at C-13 and C-17.
    c. They only differ in the substituent at C-13.
    d. They only differ in the substituent at C-17.
    e. They only differ in the substituent at C-3 and C-13.

    Answer: b Section: 9

23. Which of the following statements best describes what anabolic steroids are?

    a. They are synthetic steroids.
    b. Steroids that aid in the development of muscles.
    c. When taken in high doses, they regulate heartbeat, kidney and liver functions, and increase muscle mass.
    d. a and b
    e. a, b, and c

    Answer: d Section: 9

SHORT ANSWER

24. Palmitic acid (hexadecanoic acid) has a melting point of 63° C. Palmitoleic acid (9-hexadecenoic acid) has a melting point of 32° C. Explain this difference in melting points.

    Answer:
    Palmitoleic acid, an unsaturated fat, has a much lower melting point than the saturated palmitic acid, although they both have comparable molecular weights. This is true because the double bonds in an unsaturated fatty acid have the cis configuration. This produces a bend in the chains preventing them from packing together as tightly as saturated fats. This results in a decrease in intermolecular interactions which lowers the melting point.
    Section: 1

25. Phosphoacylglycerols form membranes by arranging themselves in a lipid bilayer. What does lipid bilayer mean?

    Answer:
    It means that the polar heads of the phosphoacylglycerols are on the outside of the bilayer, while the interior of the bilayer consists of the fatty acid chains and cholesterol.
    Section: 4

26. Give the chemical structure for sphingosine, a molecule that is found in sphingolipids.

Answer:

CH══CH─── (CH$_2$)$_{12}$CH$_3$
|
CH──OH
|
CH──NH$_2$
|
CH$_2$─OH

Section: 4

27. What is the difference between a sphingomyelin and a cerebroside? Give an example.

Answer:
A sphingomyelin is a sphingolipid that is also a phospholipid. Cerebroside is a sphingolipid that is not a phospholipid. The bottom OH in a cerebroside is bonded to a sugar residue in a β-linkage.

CH══CH─── (CH$_2$)$_{12}$CH$_3$ CH══CH─── (CH$_2$)$_{12}$CH$_3$
| |
CH── OH   CH── OH
| |
CH── NH$_2$ CH── NH$_2$
| |
CH$_2$── O── phosphate   CH$_2$── O── Sugar

Sphingomyelin   Cerebroside
(A phospholipid)   (Not a Phospholipid)

Section: 4

28. In the prostaglandin called PGF$_{2\alpha}$, what does the 2α stand for? Give an example.

Answer:
The two stands for the number of double bonds in the side chain, and the indicates that the two OH groups are on the same side of the ring.

Ex

PGF$_{2\alpha}$

Section: 5

29. Show the steps involved in the formation of prostaglandin PGE$_2$ from arachidonic acid:

Arachidonic acid

PGE$_2$

Answer:

Arachidonic acid

cyclooxygenase

hydroperoxidase

PGH$_2$

PGE$_2$

Section: 5

30. In the rod cells of the eye, retinol (vitamin A) can be oxidized and then isomerized into an aldehyde and a cis double bond at C-11 to form a molecule called 11Z-retinal. Propose a structure for 11Z-retinal.

retinol (vit. A)

Answer:

11Z-retinal

Section: 7

31. Propose a mechanism for the reaction, of dimethylallyl pyrophosphate with isopentenyl pyrophosphate to form geranyl pyrophosphate.

dimethylallyl pyrophosphate   isopentenyl pyrophosphate

geranyl pyrophosphate   +   HOPP

Answer:

geranyl pyrophosphate   +   HOPP

Section: 8

32. Draw the steroid ring system with the six-membered rings in the chair conformation.

Answer:

Section: 9

33. Show the numbering system of the steroid ring.

Answer:

Section: 9

34. Show the biosynthetic pathway in the synthesis of lanosterol (a precursor to cholesterol) from squalene.

Answer:

lanosterol

Section: 10

35. Define <u>lipid</u>

Answer:
Lipids are substances that can be extracted from cells and tissues by nonpolar organic solvents.
Section: 98

MULTIPLE CHOICE

36. Which of the following terms best describes the compound below?
$CH_3(CH_2)_{26}CO_2CH_2(CH_2)_{32}CH_3$

a. a fat
b. a wax
c. a terpene
d. an unsaturated triglyceride
e. an oil

Answer: b Section: 2

37. Which of the following terms best describes the compound below?

$$CH_2-O-\overset{\overset{\displaystyle O}{\|}}{C}-(CH_2)_{18}CH_3$$

$$CH-O-\overset{\overset{\displaystyle O}{\|}}{C}-(CH_2)_{16}CH_3$$

$$CH_2-O-\overset{\overset{\displaystyle O}{\|}}{C}-(CH_2)_{18}CH_3$$

a. a wax
b. a saturated triglyceride
c. a terpene
d. a prostaglandin
e. a lecithin

Answer: b Section: 3

38. Which of the following terms best describes the compound below?

a. a steroid
b. an unsaturated triglycaride
c. a soap
d. a complex lipid
e. a terpene

Answer: e Section: 6

39. Which of the following terms best describes the compound below?

    a. a prostaglandin
    b. a fatty acid
    c. a steroid
    d. a cephalin
    e. an essential oil

    Answer: c   Section: 9

40. Which of the following terms best describes the compound below?

    $CH_3(CH_2)_{12}CO_2H$

    a. a fatty acid
    b. an oil
    c. a wax
    d. a soap
    e. a phosphatidic acid

    Answer: a   Section: 1

41. Which of the following terms best describes the compound below?

    $CH_3(CH_2)_7CH=CH(CH_2)_7CO_2H$

    a. an unsaturated fatty acid
    b. a triglyceride
    c. a synthetic detergent
    d. a micelle
    e. isoprene

    Answer: a   Section: 1

SHORT ANSWER

42. Was the compound shown below more likely isolated from an animal or a plant? Explain.

$$CH_2-O-\overset{\overset{\textstyle O}{\|}}{C}-(CH_2)_7CH=CH(CH_2)_7CH_3$$

$$CH-O-\overset{\overset{\textstyle O}{\|}}{C}-(CH_2)_7CH=CHCH_2CH=CH(CH_2)_4CH_3$$

$$CH_2-O-\overset{\overset{\textstyle O}{\|}}{C}-(CH_2)_7CH=CH(CH_2)_7CH_3$$

Answer:
The compound shown is a polyunsaturated triglyceride. Such triglycerides are liquid at room temperature (ie, are oils) and are more likely to have been isolated from a plant.

Section: 3

MULTIPLE CHOICE

43. Which of the following terms best describes the compound below?

$$CH_2-O-\overset{\overset{\textstyle O}{\|}}{C}-(CH_2)_{14}CH_3$$

$$CH-O-\overset{\overset{\textstyle O}{\|}}{C}-(CH_2)_{14}CH_3$$

$$CH_2-O-\overset{\overset{\textstyle O}{\|}}{\underset{\underset{\textstyle O^-}{|}}{P}}-OCH_2CH_2NH_3{}^+$$

a. a phospholipid
b. a phosphogylceride
c. a cephalin
d. a molecule which contains a polar head group
e. all of the above

Answer: e Section: 4

44. Which of the following terms best describes the compound below?

    a. a lecithin
    b. a diterpene
    c. a prostaglandin
    d. a glyceride
    e. a steroid

Answer: c Section: 5

SHORT ANSWER

45. When spermaceti is heated in aqueous potassium hydroxide, $CH_3(CH_2)_{14}CH_2OH$ and $CH_3(CH_2)_{14}CO_2^- \ K^+$ are formed. Provide the structure of spermaceti.

    Answer: $CH_3(CH_2)_{14}CO_2CH_2(CH_2)_{14}CH_3$ Section: 2

46. Lauric acid has the formula $CH_3(CH_2)_{10}CO_2H$. Provide the structure of the triglyceride which was formed from three equivalents of lauric acid.

    Answer:

    Section: 3

47. How is a fat distinguished from an oil, (a) based on a physical property difference and (b) based on a difference in molecular structure?

    Answer:
    Fats have higher melting points than oils and tend to have fewer carbon-carbon double bonds (sites of unsaturation) present.
    Section: 3

48. Why and in what fashion does the presence of a double bond in a fatty acid affect its melting point relative to the system where the double bond is absent?

    Answer:
    The presence of a double bond lowers the melting point. The double bond introduces a "kink" in the carbon chain which retards efficient molecular packing and thus hinders solidification.
    Section: 3

49. What does the term <u>polyunsaturated</u> mean when used to describe a triglyceride?

    Answer:
    This term indicates the presence of several carbon-carbon double bonds in the fatty acid residues of the triglyceride.
    Section: 3

50. When a vegetable oil is partially hydrogenated, how is its structure altered and how is its melting point affected?

    Answer:
    H_2 is added across some of the carbon-carbon double bonds present. This decreases the number of sites of unsaturation and consequently increases the degree of saturation. The melting point increases.
    Section: 3

51. Draw the structure of the product which results when the compound shown below is completely hydrogenated using excess H_2.

$$CH_2-O-\overset{\overset{\displaystyle O}{\|}}{C}-(CH_2)_{10}CH=CHCH_2CH=CHCH_2CH=CHCH_2CH_3$$
$$CH-O-\overset{\overset{\displaystyle O}{\|}}{C}-(CH_2)_{10}CH=CHCH_2CH=CHCH_2CH_3$$
$$CH_2-O-\overset{\overset{\displaystyle O}{\|}}{C}-(CH_2)_{10}CH=CHCH_2CH=CH(CH_2)_4CH_3$$

Answer:

$$CH_2-O-\overset{\overset{\displaystyle O}{\|}}{C}-(CH_2)_{16}CH_3$$
$$CH-O-\overset{\overset{\displaystyle O}{\|}}{C}-(CH_2)_{16}CH_3$$
$$CH_2-O-\overset{\overset{\displaystyle O}{\|}}{C}-(CH_2)_{16}CH_3$$

Section: 3

52. Provide the structures of the major organic products which result when tristearin (shown below) is heated in an aqueous solution of sodium hydroxide.

$$CH_2-O-\overset{\overset{\displaystyle O}{\|}}{C}-(CH_2)_{16}CH_3$$
$$CH-O-\overset{\overset{\displaystyle O}{\|}}{C}-(CH_2)_{16}CH_3$$
$$CH_2-O-\overset{\overset{\displaystyle O}{\|}}{C}-(CH_2)_{16}CH_3$$

Answer: $HOCH_2CHOHCH_2OH$ and $CH_3(CH_2)_{16}CO_2^-Na^+$ Section: 3

54. Provide a sketch of a lipid bilayer. Label the polar and nonpolar regions.

Answer:

Section: 4

MULTIPLE CHOICE

55. Most of the principal sex hormones in humans are:

   a. triglycerides.
   b. prostaglandins.
   c. steroids.
   d. phospholipids.
   e. terpenes.

Answer: c Section: 9

SHORT ANSWER

56. Draw a neat conformational structure of the compound below.

Answer:

Section: 9

57. How would you prepare $CH_3(CH_2)_{14}CH_2NH_2$ from palmitic acid, $CH_3(CH_2)_{14}CO_2H$?

Answer:
1. $SOCl_2$
2. NH_3
3. $LiAlH_4$

Section: 98

58. A terpene is believed to have one of the structures below. Based on the isoprene rule, which structure is more likely. Explain your reasoning.

Answer:
This compound contains an integral number of isoprene units.

Section: 6

59. Circle the isoprene units present in menthol below.

Answer:

Section: 6

60. Would the addition of HCl to the double bond of oleic acid, $CH_3(CH_2)_7CH=CH(CH_2)_7CO_2H$, occur with high regioselectivity? Explain.

Answer:
No. The two possible carbocation intermediates are very similar in structure and stability and would thus be formed at comparable rates.
Section: 98

61. What organic compounds are formed when linoleic acid, $CH_3(CH_2)_4CH=CHCH_2CH=CH(CH_2)_7CO_2H$, is ozonolyzed and subsquently treated with H_2O_2?

Answer: $CH_3(CH_2)_4CO_2H$, $HO_2CCH_2CO_2H$, and $HO_2C(CH_2)_7CO_2H$ Section: 1

62. To what common biological uses are waxes put?

Answer:
The feathers of birds are coated with wax to make them water-repellent. Bees use wax as a structural element of their hives, while other insects secrete waxes to waterproof their exoskeletons. Some vertebrates secrete wax to lubricate their fur and to render it water-repellent. Plants also use waxes to protect their leaves from parasites and to minimize water evaporation.
Section: 2

MULTIPLE CHOICE

63. When glycerol is mixed with excess lauroyl chloride and excess myristoyl chloride in the same reaction vessel, how many distinct triacylglycerols can result?

    a. 2
    b. 3
    c. 4
    d. 6
    e. 8

    Answer: e   Section: 3

SHORT ANSWER

64. How is Olestra made and how does it act as a fat substitute?

    Answer:
    Olestra is made by esterifying all the OH groups of sucrose with fatty acids obtained from cottonseed oil and soybean oil. Therefore its component parts are table sugar and vegetable oil. Olestra works as a fat substitute because its ester linkages are too hindered to be hydrolyzed by digestive enzymes. As a result, Olestra tastes like fat, but has no caloric value since it cannot be digested.
    Section: 3

65. Why are polyunsaturated fats and oils more easily oxidized than saturated ones?

    Answer:
    Triacylglycerols are oxidized by oxygen via a radical chain reaction. The polyunsaturated compounds are more easily oxidized since hydrogen atom abstraction in the radical chain process results in resonance-stabilized allylic radicals.
    Section: 3

66. Vitamin E can protect cell membranes from degradation. Through what mechanism is such degradation occurring, and how does vitamin E inhibit this mechanism of degradation?

Answer:
The unsaturated fatty acid chains of phosphoacylglycerols are susceptible to reaction with oxygen through a radical chain process. Since such phosphoacylglycerols are large components of cell membranes, this oxidation can lead to membrane degradation. Vitamin E protects fatty acid chains from this oxidative pathway because it reacts more rapidly with oxygen than the chains do. Vitamin E thus functions as an antioxidant.

Section: 4

MULTIPLE CHOICE

67. Which of the following statements concerning leukotrienes is (are) correct?

   a. Leukotrienes can be produced directly from 2 moles of myristic acid.
   b. Leukotrienes contain 5 conjugated double bonds.
   c. Leukotrienes are implicated in allergic reactions, inflammatory reactions, and anaphalytic shock.
   d. Both a and c.
   e. All of the above.

Answer: c   Section: 5

68. What is the IUPAC name for isoprene?

   a. 1,3-pentadiene
   b. 2,4-pentadiene
   c. 2,3-dimethyl-1,3-butadiene
   d. 2-methyl-1,3-butadiene
   e. none of the above

Answer: d   Section: 6

SHORT ANSWER

69. Provide the structure of α-farnesene, a simple, acyclic sesquiterpene.

Answer:

Section: 6

MULTIPLE CHOICE

70. Which of the following is not an intermediate in the biosynthesis of isopentenyl pyrophosphate?

    a. hydroxymethylglutaryl-CoA
    b. mevalonylphosphate
    c. acetoacetyl-CoA
    d. dimethylallyl phosphate
    e. mevalonic acid

    Answer: d   Section: 8

Chapter 25:   Nucleosides, Nucleotides, and Nucleic Acids

MULTIPLE CHOICE

1. Which of the following statements explains what nucleic acids are?

   a. They are acidic in nature.
   b. They are the carriers of genetic information.
   c. DNA and RNA are collectively known as nucleic acids.
   d. Nucleic acids are chains of five-membered ring sugars, each bonded to a heterocyclic amine, linked by phosphate groups.
   e. All of the above.

   Answer: e

2. Which of the following phosphate groups is found in nucleic acids?

   Answer: c Section: 7

3. Which of the following are known as pyrimidines?

    a. Cytosine and thymine
    b. Adenine and cytosine
    c. Adenine and thymine
    d. Guanine and thymine
    e. Adenine and guanine

Answer: a Section: 1

4. Which of the following bases distinguish DNA from RNA?

    a. Adenine and cytosine
    b. Guanine and uracil
    c. Thymine and uracil
    d. Uracil and guanine
    e. Adenine and thymine

Answer: c Section: 1

5. Which of the following is the proper name of the nucleoside made from adenine and found in DNA?

    a. adenosine
    b. 2'-deoxyadenosine
    c. 2-deoxyadenosine
    d. 2'-oxyadenosine
    e. 2'-deoxyadenine

Answer: b Section: 1

6. For what compound is ATP an abbreviation?

    a. Adenine triphosphate
    b. Adenosine tetraphosphate
    c. Adenine tetraphosphate
    d. Adenosine triphosphate
    e. Adenine triphosphoric acid

Answer: d Section: 2

7. When phosphoric acid is heated it loses water to form an anhydride known as pyrophosphate or pyrophosphoric acid. What is the structure of pyrophosphoric acid?

Answer: b   Section: 1

8. Which of the following statements best explains the source of energy behind the ATP molecule?

    a. The energy released when a phosphoanhydride bond of ATP is broken
    b. The energy released when adenosine binds to the phosphate
    c. The energy released when ribose binds to the phosphate
    d. The energy released when adenine binds to ribose
    e. The energy released when ATP is reduced

Answer: a Section: 2

9. Which of the following is the first step in the mechanism of the reaction shown below?

$$CH_3-\overset{\overset{\displaystyle O}{\|}}{C}-\overset{\ominus}{O} + R-SH + ATP \longrightarrow CH_3-\overset{\overset{\displaystyle O}{\|}}{C}-SR + ADP + \overset{\ominus}{O}-\overset{\overset{\displaystyle O}{\|}}{\underset{\underset{\displaystyle O}{\ominus|}}{P}}-OH$$

a. Carboxylate ion attacks the thiol.
b. Carboxylate ion attacks the α-phosphorus of ATP.
c. Carboxylate ion attacks the γ-phosphorus of ATP.
d. Thiol molecule attacks the β-phosphorus of ATP.
e. Thiol molecule attacks the acyl group.

Answer: c   Section: 3

10. Which of the following reactions converts nucleotide subunits into nucleic acids?

a. Nucleophilic attack on the α-phosphorus of ATP
b. Nucleophilic attack on the β-phosphorus of ATP
c. Nucleophilic attack on the γ-phosphorus of ATP
d. a and b
e. a, b and c

Answer: a   Section: 3

11. Which of the following metal ions complexes with cellular ATP to neutralize some of its negative charges at the active site of an enzyme in order that nucleophilic substitution reactions proceed rapidly?

a. Manganese
b. Iron
c. Cobalt
d. Magnesium
e. Calcium

Answer: d   Section: 3

12. Which of the following reaction sequences best describes the formation of the nucleic acid in the biosynthesis of DNA?

    a. Nucleophilic attack of a 5'-OH group of one nucleotide on the β -phosphorus of another
    b. Nucleophilic attack of a 5'-OH group of one nucleotide on the α -phosphorus of another
    c. Nucleophilic attack of a 3'-OH group of a nucleotide on the γ -phosphorus of another
    d. Nucleophilic attack of a 3'-OH group of one nucleotide on the β -phosphoros of another
    e. Nucleophilic attack of a 3'-OH group of one nucleotide on the α -phosphoros of another

    Answer: e Section: 7

13. Which of the following constitutes the primary structure of a nucleic acid?

    a. The sequence of the bases in the strand
    b. The sequence of the riboses in the strand
    c. The sequence of the deoxyriboses in the strand
    d. The sequence of the phosphodiesters in the strand
    e. The sequence of the 5'-OH groups in the strand

    Answer: a Section: 7

14. Which of the following statements is <u>not</u> true about the structure of DNA as proposed by Watson and Crick?

    a. The number of adenines in DNA is equal to the number of thymines.
    b. The number of cytosines is equal to the number of guanines.
    c. DNA consists of two strands of nucleic acids with the sugar-phosphate backbone on the inside and the bases on the outside.
    d. The chains in DNA are held together by hydrogen bonding.
    e. Adenine always pairs with thymine and guanine always pairs with cytosine.

    Answer: c Section: 7

15. How many hydrogen bonds exist between the cytosine/guanine base pairs?

    a. 1
    b. 2
    c. 3
    d. 4
    e. 5

    Answer: c Section: 7

16. Which of the following is the conventional method of writing the base sequence of a portion of a DNA chain?

    a. 5'-end--bases--5'-end
    b. 5'-end--bases--3'-end
    c. 3'-end--bases--5'-end
    d. 3'-end--bases--3'-end
    e. 5'-end--bases--2'-end

    Answer: b Section: 7

17. In which of the following helical forms does DNA in living organisms mainly exist?

    a. Z-helix
    b. A-helix
    c. B-helix
    d. b and c
    e. a, b and c

    Answer: c Section: 8

18. How many billion base pairs are there in the total DNA of a human cell?

    a. 3
    b. 5
    c. 8
    d. 11
    e. 13

    Answer: a Section: 9

19. Which of the following statements best describes the meaning of transcription?

    a. The synthesis of rRNA from a DNA blueprint
    b. Transcription is synonymous with replication
    c. The synthesis of identical copies from a DNA blueprint
    d. The synthesis of mRNA from a DNA blueprint
    e. The synthesis of tRNA from a DNA blueprint

    Answer: d Section: 10

20. In an entire DNA strand, what percentage consists of genetic information and what percentage consists of introns?

    a. 90% genetic information, 10% introns
    b. 10% genetic information, 90% introns
    c. 50% genetic information, 50% introns
    d. 100% genetic information, 0% introns
    e. 99% genetic information, 1% introns

    Answer: b Section: 10

21. Which of the following statements is <u>not</u> true about RNA?

    a. It is single stranded.
    b. It is smaller than DNA.
    c. It has fewer base pairs than DNA.
    d. It exists exclusively in the nucleus of the cell.
    e. There are three kinds: rRNA, mRNA, and tRNA.

    Answer: d Section: 10

22. What is the composition of a ribosome?

    a. 40% protein, 60% rRNA
    b. 60% protein, 40% tRNA
    c. 40% protein, 60% tRNA
    d. 40% protein, 60% mRNA
    e. 100% protein, 0% rRNA

    Answer: a Section: 11

23. Which of the following is true about tRNA?

    a. All tRNA's have a CCG sequence at the 3'-end.
    b. All tRNA's have a CCG sequence at the 5'-end.
    c. All tRNA's have a CCA sequence at the 5'-end.
    d. All tRNA's have a CCA sequence at the 3'-end.
    e. All tRNA's have a CAC sequence at the 3'-end.

    Answer: d Section: 12

24. Which of the following base sequences would most likely be recognized by a restriction endonuclease?

    a. ACGCGT
    b. ACGGGT
    c. ACACGT
    d. ACATCGT
    e. CCAACC

    Answer: a Section: 15

25. Which of the following is a phosphoramidite?

a.
$$
\begin{array}{c}
\text{O} \\
\parallel \\
\text{RO} \quad \text{CNH}_2 \\
\diagdown \quad \diagup \\
\text{P} \\
\mid \\
\text{OR}
\end{array}
$$

b.
$$
\begin{array}{c}
\text{NH}_2 \diagdown \\
\text{P}-\text{NH}_2 \\
\diagup \\
\text{H}_2\text{N}
\end{array}
$$

c.
$$
\begin{array}{c}
\text{NR}_2 \\
\mid \\
\text{P}-\text{OR} \\
\diagup \\
\text{RO}
\end{array}
$$

d.
$$
\begin{array}{c}
\text{OR} \\
\mid \\
\text{P} \\
\diagup \ \diagdown \\
\text{RO} \quad \text{OR}
\end{array}
$$

e.
$$
\begin{array}{c}
\text{NR}_2 \\
\mid \\
\text{P}-\text{NR}_2 \\
\mid \\
\text{NR}_2
\end{array}
$$

Answer: c Section: 16

26. A new method, called the H-phosphonate method, is replacing the phosphoramidite method for preparing oligonucleotides. Which of the following is the first step in the H-phosphonate method?

a. Oxidation with iodine.
b. The 5'-OH group of the nucleoside reacts with the anhydroside.
c. The protecting group is removed by ammonium hydroxide.
d. The DMTr group is removed with mild acid.
e. The monomer, H-phosphonate, is activated by reacting it with acyl chloride to form an anhydride.

Answer: e Section: 16

27. Which of the following is <u>not</u> true about retroviruses?

   a. The genetic information is contained in RNA.
   b. The genetic information flows from DNA to RNA.
   c. Drugs have been designed to interfere with the synthesis of DNA by retroviruses.
   d. AZT is a design drug that was synthesized to combat the HIV retrovirus.
   e. The retrovirus uses the sequence of bases in RNA as a template to synthesize DNA which enters and infects the host cell.

Answer: b Section: 17

SHORT ANSWER

28. What is a nucleoside? Give an example.

Answer: A nucleoside is a base plus a sugar.

Ex.

Cytidine 5'-monophosphate

Section: 1

29. What is a nucleotide? Give an example.

Answer:
A nucleotide is a nucleoside with one of its OH groups bonded to phosphoric acid in an ester linkage. Nucleotides of RNA are called ribonucleotides while the nucleotides of DNA are called deoxyribonucleotides.

Ex.

Cytidine 5'-monophosphate

Section: 1

30. Draw the chemical structure of adenosine triphosphate, (ATP).

Answer:

Section: 2

31. Propose a mechanism for the conversion of D-glucose to D-glucose-6-phosphate using ATP as a source of the phosphate.

Answer:

D-Glucose-6-phosphate

Section: 2

603

32. Propose a mechanism for the following reaction:

$$CH_3-C(=O)-O^{\ominus} + RSH + ATP \longrightarrow CH_3-C(=O)-SR + ADP + {}^{\ominus}O-P(=O)(O^{\ominus})-OH$$

Answer:

Section: 3

33. A commonly known nucleotide is adenosine 3',5'-phosphate (cAMP). What is the structure of cAMP?

Answer:

cAMP

Section: 6

34. Propose a mechanism for the formation of cyclic AMP from ATP.

Answer:

Section: 6

35. What are stacking interactions in DNA and what purpose do they serve?

    Answer:
    Stacking interactions are weak attractive forces (van der Waal's forces) that contribute significantly to the stability of the double helix. Stacking interactions are strongest between two purines and weakest between two pyrimidines.
    Section: 7

36. What is meant by semiconservative DNA replication?

    Answer:
    When DNA strands are replicated, two daughter molecules of DNA are produced. Each daughter molecule contains one of two original strands plus a newly synthesized strand. This is called semiconservative DNA replication.
    Section: 7

37. Assuming that the human genome with its 3 billion base pairs is entirely in a B-helix, how long is the DNA in a human cell?

    Answer:
    Recall that B-DNA has 10 base pairs and a rise of 3.3 angstrom. Therefore, the length of the DNA would be:

    $$\frac{3 \times 10^9 \text{ base pairs}}{10 \text{ base pairs}} \times 3.3 \text{ angstrom} \times 10^{-8} \frac{\text{cm}}{\text{angstrom}} = 9.9 \text{ cm}$$

    Section: 8

38. What is a codon? Give an example.

    Answer:
    A codon is a sequence of three bases that specifies a particular amino acid that is to be incorporated into a protein.
    Ex. CUU is the code for leucine.
    Section: 13

39. Methionine is the first amino acid incorporated into the heptapeptide specified by the following mRNA:
    5'—AUGGACCCCGUUAUUAAACAC-3'. What is the sequence of the amino acid in the heptapeptide?

    Answer:
    Using the genetic code table, the sequence is
    Met-Asp-Pro-Val-Ile-Lys-His.
    Section: 13

605

40. What are Maxam-Gilbert and Sanger sequencing?

    Answer: They are methods used to sequence restriction fragments of DNA.

    Section: 15

41. What are antigene and antisense agents?

    Answer:
    They are synthetic polymers that bind to specific sequences of a nucleic acid strand. When the polymer binds to DNA, it is called an antigene agent, and when the polymer binds to RNA, it is called an antisense agent.
    Section: 17

42. What are the four common ribonucleosides?

    Answer: cytidine, uridine, adenosine, and guanosine   Section: 1

43. What is the major structural difference between a ribonucleoside and a ribonucleotide?

    Answer:
    A ribonucleotide is a ribonucleoside which has been phosphorylated at its 5' carbon
    Section: 1

44. Besides a possible difference in base structure, what is the major structural difference between ribo- and deoxyribonucleosides?

    Answer:
    Deoxyribonucleosides have a hydrogen instead of a hydroxyl group at the C2 position of the sugar.
    Section: 1

45. Name the two pyrimidine bases which serve as components in deoxyribonucleotides.

    Answer: cytosine and thymine   Section: 1

46. Name the two purine bases which serve as components in ribonucleosides.

    Answer: adenine and guanine   Section: 1

47. Show the hydrogen bonding which occurs when guanine and cytosine form a base pair.

    Answer:

    Section: 7

48. Show the hydrogen bonding which occurs when adenine and thymine form a base pair.

    Answer:

    Section: 7

49. Show that you understand how the structural components of DNA are bonded together by drawing a linear segment that contains 3 base residues.

    Answer: Other choices for the base components are possible.

    Section: 1

MULTIPLE CHOICE

50. Which of the following statements concerning DNA, RNA, and their components is (are) not correct?

    a. Phosphoric acid has three dissociable OH groups which can also react with alcohols to form phosphoesters.
    b. Across the DNA helix purine bases hydrogen bond only to other purine bases.
    c. The two substituted pyrimidines in DNA are cytosine and thymine.
    d. Both b and c.
    e. None of the above.

    Answer: b   Section: 1

51. Identify the nucleoside shown below.

    a. adenosine
    b. guanosine
    c. cytidine
    d. uridine
    e. thymidine

    Answer: b   Section: 1

SHORT ANSWER

52. Draw the structure of 2'-deoxycytidine 3'-monophosphate.

Answer:

Section: 1

53. Draw the structure of UDP.

Answer:

Section: 1

54. Why is the ΔG° ' value for the hydrolysis of ATP large and negative? Give three reasons.

Answer:
1. Because of ATP's greater negative charge, more electrostatic repulsions are present in ATP than in ADP or phosphate. These repulsions destabilize ATP relative to ADP and phosphate.
2. Because ATP has 3.3 negative charges at physiological pH, while the sum on ADP and phosphate is 4.4, there is more solvation in the products than in the reactant. This stabilizes the products relative to the reactants.
3. There is greater resonance stabilization in the product anions (ADP and phosphate) than in ATP.

Section: 4

MULTIPLE CHOICE

55. Which of the following explains the kinetic stability of ATP?

a. The hydrolysis of ATP occurs rapidly in the absence of a catalyst because the phosphate bonds are thermodynamically labile.
b. ATP is positively charged at physiological pH.
c. Water, the major component of cytoplasm, is a strong and very reactive nucleophile.
d. ATP is negatively charged at physiological pH and thus repels the approach of nucleophiles.
e. None of the above.

Answer: d   Section: 5

56. Which of the following statements concerning nucleic acids is not correct?

a. The enzymes that catalyze DNA synthesis are known as DNA polymerases.
b. The growing DNA polymer is synthesized in the 5' to 3' direction.
c. RNA strands are synthesized in the 3' to 5' direction.
d. In DNA the distance between the strands of the double helix is relatively constant.
e. Unlike DNA, RNA is easily cleaved because the 2'-OH group of the ribose can act as the nucleophile that cleaves the strand.

Answer: c   Section: 7

57. What is the sequence of bases in the strand of DNA that is complementary to that shown below?
5'-T-T-C-A-C-T-G-A-C-3'

  a. 5'-T-T-C-A-C-T-G-A-C-3'
  b. 3'-T-T-C-A-C-T-G-A-C-5'
  c. 5'-A-A-G-T-G-A-C-T-G-3'
  d. 5'-G-T-C-A-G-T-G-A-A-3'
  e. 3'-U-U-C-A-C-U-G-A-C-5'

Answer: d   Section: 7

Chapter 26: Synthetic Polymers

MULTIPLE CHOICE

1. The joining together of many monomers to form a very large molecule is called?

   a. Polymerization
   b. Monomerization
   c. Dimerization
   d. Hydroxylation
   e. Homogenization

   Answer: a Section: 98

2. The first polymer, invented by A. Parke in 1856, was

   a. Polyethlene
   b. Polyester
   c. Polystyrene
   d. Celluloid
   e. Teflon

   Answer: d Section: 98

3. What is the structure of the monomer from which the following polymer is made?

   $$-CH-CH-CH-CH-CH-CH-CH-CH-$$
   $$\underset{CH_3}{|}\;\underset{CH_3}{|}\;\underset{CH_3}{|}\;\underset{CH_3}{|}\;\underset{CH_3}{|}\;\underset{CH_3}{|}\;\underset{CH_3}{|}\;\underset{CH_3}{|}$$

   a. CH_3CH_3
   b. $CH_2=CH_2$
   c. $CH_3CH=CH_2$
   d. $CH_2=CHCH=CH_2$
   e. $CH_3CH=CHCH_3$

   Answer: e Section: 2

4. What is the structure of the monomer from which the following polymer is made?

   a. I
   b. II
   c. III
   d. IV
   e. V

—CHCH$_2$–CHCH$_2$–CHCH$_2$–CHCH$_2$—

I. CH$_2$CH$_3$

II. CH=CH$_2$

III. CH$_2$CH$_2$CH$_3$

IV. CH=CHCH$_3$

V. CH=CHCH=CH$_2$

Answer: b Section: 2

5. Which of the following addition polymers results from the reaction below?

$$nCF_2{=}CF_2 \xrightarrow{\text{catalyst}} ?$$

   a. [=CF–CF=]n
   b. [CF$_3$-CF$_3$]n
   c. [-CF$_2$-CH=CHCF$_2$-]n
   d. [-CF$_2$-CF$_2$-]n
   e. [-CF$_2$=CF$_2$-]n

Answer: d Section: 2

6. Which dimer (step-growth dimerization) will form from the following monomer?

$$\begin{array}{c} O \\ \parallel \\ 2 \ CH_2 - C - OH \\ \mid \\ NH_2 \end{array} \longrightarrow$$

a.
$$\begin{array}{c} O \ O \\ \parallel \ \parallel \\ CH_2COCCH_2 \\ \mid \qquad \mid \\ NH_2 \quad NH_2 \end{array}$$

b.
$$\begin{array}{c} O \quad O \\ \parallel \quad \parallel \\ CH_2CONHCOH \\ \mid \\ NH_2 \end{array}$$

c.
$$\begin{array}{c} O \qquad O \\ \parallel \qquad \parallel \\ CH_2CNHCH_2COH \\ \mid \\ NH_2 \end{array}$$

d.
$$\begin{array}{c} O \qquad\qquad O \\ \parallel \qquad\qquad \parallel \\ HOCCH_2NHNHCH_2COH \end{array}$$

e.
$$\begin{array}{c} O \qquad\qquad O \\ \parallel \qquad\qquad \parallel \\ HOCCH_2N = NCH_2COH \end{array}$$

Answer: c   Section: 6

7. Which of the following statements best describes what <u>alpha olefins</u> are?

a. Polysubstituted ethylenes
b. Monosubstituted ethylenes
c. Disubstituted ethylenes
d. Ethylenes
e. Isoprenes

Answer: b   Section: 2

8. The choice of mechanism in chain-growth polymerization depends on which of the following?

   a. Structure of the monomer and initiator used
   b. Structure of the polymer and initiator used
   c. Temperature and pressure of the reaction
   d. a and b
   e. a, b, and c

   Answer: a Section: 2

9. Chain-growth polymerization proceeds by which of the following mechanisms?

   a. Radical polymerization
   b. Cationic polymerization
   c. Anionic polymerization
   d. a and b
   e. a, b, and c

   Answer: e Section: 2

10. Which of the following species can best serve as a radical initiator for radical polymerization?

    a. ROH
    b. ROR
    c. ROOR
    d. RCOOR
    e. RCOOH

    Answer: c Section: 2

11. Which of the following is the first chain propagating step in the radical polymerization of ethylene?

    a. $ROOR + CH_2{=}CH_2 \longrightarrow ROCH_2CH_2OR$

    b.
    $RO{\cdot} + CH_2{=}CH_2 \longrightarrow ROCH_2\overset{\cdot}{C}H_2$

    c. $RO{\cdot} + CH_2{=}CH_2 \longrightarrow ROCH{=}CH_2 + H{\cdot}$

    d.
    $ROOR + CH_2{=}CH_2 \longrightarrow 2\ ROH + \overset{\cdot}{C}H{=}\overset{\cdot}{C}H$

    e.
    $2RO{\cdot} + CH_2{=}CH_2 \longrightarrow 2\ ROH + \overset{\cdot}{C}H{=}\overset{\cdot}{C}H$

    Answer: b Section: 2

12. Which of the following is a possible step to terminate a chain in the radical polymerization of ethylene?

    a. $ROOR + CH_2{=}CH_2 \longrightarrow ROCH_2CH_2OR$

    b. $CH_2{=}CH_2 + CH_2{=}CH_2 \longrightarrow CH_3CH{=}CHCH_3$

    c.
    $RO\cdot + CH_2{=}CH_2 \longrightarrow RO\overset{\bullet}{C}H_2{=}CH_2$

    d.
    $RO CH_2\overset{\bullet}{C}H_2 + CH_2{=}CH_2 \longrightarrow ROCH_2CH_2CH_2\overset{\bullet}{C}H_2$

    e.
    $R_2{-}[CH_2CH_{2\,n}]{-}CH_2\overset{\bullet}{C}H_2 \longrightarrow$

    $RO{-}[CH_2CH_2]{-}CH_2CH_2CH_2CH_2{-}[CH_2CH_2]{-}OR$

    Answer: e   Section: 2

13. Which of the following statements best explains the process known as <u>chain transfer</u>?

    a. A process that can control the rate of production of the polymer
    b. A process that can control the molecular weight of the polymer
    c. A process that can control the reactivity of the monomer
    d. A process that can slow down the polymerization reaction
    e. A process that can inhibit the polymerization reaction

    Answer: b   Section: 2

14. Chain-growth polymerization exhibits a marked preference for:

    a. tail-to-tail addition
    b. head-to-head addition
    c. head-to-tail addition
    d. head-to-middle addition
    e. middle-to-tail addition

    Answer: c   Section: 2

15. Which of the following are characteristic of <u>linear</u> polyethylene?

    a. They are branched polymers.
    b. They are known as high-density polyethylene.
    c. They are soft and flexible polymers.
    d. a and b.
    e. a and c.

    Answer: b   Section: 2

16. Which of the following species can best serve as an initiator for cationic polymerization?

    a. ROOR
    b. ROH
    c. ROR
    d. $AlCl_3$
    e. RCOOR

    Answer: d Section: 2

17. Which of the following is the chain-initiating step in the cationic polymerization of propene?

    a. I
    b. II
    c. III
    d. IV
    e. V

    I. BF_3 + CH_2=CH $\longrightarrow$ F_2BCH=CH + HF
       | |
       CH_3 CH_3

    II. BF_3 + CH_2=CH $\longrightarrow$ CH_2=CBF_2 + HF
        | |
        CH_3 CH_3

    III. BF_3 + CH_2=CH $\longrightarrow$ CH_2=CH + HF
         | |
         CH_3 CH_2BF_2

    IV. BF_3 + CH_2=CH $\longrightarrow$ $\overset{\oplus}{CH_2}CH\overset{\ominus}{BF_3}$
        | |
        CH_3 CH_3

    V. BF_3 + CH_2=CH $\longrightarrow$ $\overset{\ominus}{F_3}BCH_2\overset{\oplus}{CH}$
       | |
       CH_3 CH_3

    Answer: e Section: 2

18. Which of the following is more likely to undergo both radical <u>and</u> cationic polymerization?

    a. I
    b. II
    c. III
    d. IV
    e. V

    I. $CH_2=CH_2$

    II. $C_6H_5-CH=CH_2$

    III. $CH_2=CHCl$

    IV. $CH_2=CHC\equiv N$

    V. $CH_2=C(CH_3)COOR$

Answer: b Section: 2

19. Which of the following monomers has the greatest ability to undergo cationic polymerization?

    a. I
    b. II
    c. III
    d. IV
    e. V

I.

$CH=CH_2$

II.

$CH=CH_2$

OCH_3

III.

$CH=CH_2$

NH_2

IV.

$CH=CH_2$

CH_3

V.

$CH=CH_2$

O_2N

Answer: c Section: 2

20. Which of the following monomers has the greatest ability to undergo anionic polymerization?

    a. I
    b. II
    c. III
    d. IV
    e. V

    I.

    II.

    III.

    IV.

    V.

    Answer: e   Section: 2

21. Which of the following species can best serve as an initiator for anionic polymerization?

    a. $CH_3CH_2CH_2CH_2Li$
    b. BF_3
    c. HOOH
    d. $CH_3CH_2CH_2CH_2OH$
    e. $CH_3CH_2CH_2CH_2OCH_2CH_2CH_2CH_3$

    Answer: a   Section: 2

22. Which of the following is more likely to form non-terminated polymeric chains called <u>living polymers</u>?

    a. Cationic polymerization
    b. Free radical polymerization
    c. Anionic polymerization
    d. a and b
    e. a and c

    Answer: c   Section: 2

23. A substituted ethylene polymer in which the substituents are randomly oriented is called:

    a. A Ditactic polymer
    b. An Eutatic polymer
    c. An Isotactic polymer
    d. A Syndiotatic polymer
    e. An Atactic polymer

    Answer: e Section: 3

24. Which of the following polymers is arranged in an <u>isotactic</u> configuration?

    a. I
    b. II
    c. III
    d. IV
    e. V

    I.   H   H   CH₃
         H₃C⋯H₃C⋯H⋯

    II. H   CH₃ CH₃
         H₃C⋯H⋯H⋯

    III. H   CH₃ H
         H₃C⋯H⋯H₃C⋯

    IV. H   H   H
         H₃C⋯H₃C⋯H₃C⋯

    V.   H   H   H
         H₃C⋯H₃C⋯H⋯
             CH₃ CH₃
             H   H

    Answer: d Section: 3

25. Which of the following initiators are known as <u>Ziegler-Natta catalysts</u>, which were discovered in 1953 in order to control the stereochemistry of polymers?

    a. Al/Ti initiators
    b. Ni/Cd initiators
    c. Zn/Hg initiators
    d. Pb/C initiators
    e. Pb/PbO₂ initiators

    Answer: a Section: 3

26. What is the difference between natural rubber and Gutta-percha rubber?

    a. Natural rubber is hard and brittle while gutta-percha is soft and rubbery.
    b. Natural rubber is natural while gutta-percha is synthetic rubber.
    c. Natural rubber has a cis configuration while gutta-percha has a trans configuration.
    d. Natural rubber and gutta-percha rubber are enantiomers.
    e. The monomer for natural rubber is isoprene while the monomer for gutta-percha is 1,3-butadiene.

    Answer: c Section: 4

27. Which of the following statements best describes what vulcanization means?

    a. A process by which rigid and non-sticky rubber can be made soft and sticky
    b. A chemical process discovered by Charles Schultz in 1944
    c. The heating of rubber with sulfur
    d. A chemical process that destroys cross-linking in rubber
    e. All of the above

    Answer: c Section: 4

28. Which of the following are copolymers?

    a. An Alternating copolymer
    b. A Block copolymer
    c. A Random copolymer
    d. A Graft copolymer
    e. All of the above

    Answer: e Section: 5

29. Which of the following is a step-growth polymer?

    a. Polyethylene
    b. Polyester
    c. Polypropylene
    d. Polystyrene
    e. Plexiglass

    Answer: b Section: 6

30. What are <u>aramides</u>?

    a. Aromatic polyester polymers
    b. Aromatic amine polymers
    c. Alkyl halide polymers
    d. Aromatic polyamide polymers
    e. Aliphatic polyamide polymers

    Answer: d Section: 6

31. What monomers formed the following polymer?

    a. I
    b. II
    c. III
    d. IV
    e. V

    Answer: a Section: 6

32. Which of the following are characteristics of thermosetting polymers?

    a. Strong and rigid
    b. High degree of cross-linking
    c. Can be remelted by heating
    d. a and b
    e. a, b, and c

    Answer: d Section: 7

33. What are <u>elastomers</u>?

    a. Elastic organic molecules that are added to polymers to make them harder
    b. Plastics that stretch and then revert to their original shape
    c. Polymers that are soft and sticky
    d. Plastics that are hard and brittle
    e. Elastic molecules that are added to polymers to make them stretch more

    Answer: b Section: 7

34. Which of the following is characteristic of oriented polymers?

    a. Weaker than steel
    b. Chains are arranged in parallel fashion
    c. Chains are arranged in perpendicular fashion
    d. Polymer chains are made from sulfur atoms
    e. All of the above

    Answer: b Section: 7

SHORT ANSWER

35. List the two main groups of polymers and give one example of each.

    Answer:
    Polymers are divided into two main groups:
    a. Biopolymers- ex. DNA
    b. Synthetic polymers- ex. Polyester
    Section: 98

36. List the two major classes of synthetic polymers and give one example of each.

    Answer:
    a. Addition polymers, also known as chain-growth polymers- ex. Polystyrene
    b. Condensation polymers, also known as step-growth polymers- ex. Nylon
    Section: 1

37. List the three phases in the mechanism of chain-growth polymerization.

    Answer:
    1. Initiation
    2. Propagation
    3. Termination
    Section: 2

38. Show the three methods by which the following radical can be terminated.

$$RO-CH_2-\overset{\bullet}{C}H_2$$

Answer:

a. $2 \ RO-CH_2\overset{\bullet}{C}H_2 \longrightarrow RO-CH_2CH_2CH_2CH_2-OR$

b. $2 \ RO-CH_2\overset{\bullet}{C}H_2 \longrightarrow RO-CH=CH_2 + ROCH_2CH_3$

c. $RO-CH_2\overset{\bullet}{C}H_2 + impurity \longrightarrow RO-CH_2CH_2-impurity$

Section: 2

39. Explain why some monomers are better able to undergo chain-growth polymerization by a radical mechanism.

Answer:
A monomer with a substituent (X) that is able to stabilize the growing radical species by resonance will undergo chain-growth polymerization by a free radical mechanism.

Section: 2

40. What are common factors that enter into the choice of radical initiators for chain-growth polymerization?

Answer:
1. A relatively weak oxygen-oxygen bond
2. Desired solubility of the initiator
3. Temperature of the polymerization reaction

Section: 2

41. Show the mechanism for the formation of a segment of polystyrene containing two molecules of styrene and initiated by hydrogen peroxide.

Answer:

Section: 2

42. When 3-methyl-1-butene

$$CH_2=CHCH(CH_3)_2$$

undergoes cationic polymerization, the following expected carbocation is formed.

$$X-CH_2-\overset{\oplus}{CH}-CH(CH_3)_2$$

However, the following carbocation is also found in the reaction.

$$X-CH_2-CH_2-\overset{\oplus}{C}(CH_3)_2$$

Explain.

Answer:
The expected carbocation is a 2° carbocation which may rearrange to a 3° carbocation, a more stable carbocation. This is a classic 1,2-hydride shift.

Section: 2

43. Propose a mechanism for the following polymerization reaction:

Answer:

Section: 6

44. When 2,2-dimethyloxirane (see figure below) undergoes anionic polymerization, the nucleophile attacks the least substituted carbon of the epoxide, but when it undergoes cationic polymerization, the nucleophile attacks the most substituted carbon of the epoxide. Explain why this is so.

Answer:
In anionic polymerization, the nucleophile will seek the least <u>hindered</u> carbon similar to an S_N2 reaction. However, in cationic polymerization, the reaction involves a partial carbocation and the attack is on the most hindered (most stable) <u>carbocation</u>.

Section: 2

45. What are <u>crystallites</u>?

Answer:
The regions of polymers in which the chains are highly ordered with respect to one another.

Section: 7

46. Describe what thermoplastic polymers are.

Answer:
Polymers that have both ordered crystalline regions and amorphous, non-crystalline regions. The polymers are hard at room temperature but can be remelted and remolded by heating. Thermoplastic polymers are the plastics that we most often encounter in our daily lives.

Section: 7

47. Describe what plasticizers are and give an example.

Answer:
An organic compound that dissolves in the polymer and allows the polymer chains to slide by each other.

Ex. Dibutyl phthalate,

Section: 7

48. What are biodegradable polymers? Give an example.

Answer:
Polymers that can be broken into smaller segments by enzyme-catalyzed reaction using enzymes produced by microorganisms.
Ex. Inserting hydrolyzable esters into polymers can cause weak links in the polymer which are hydrolyzable by enzyme-catalyzed reactions.
Section: 7

49. Show the reaction which occurs when benzoyl peroxide is heated.

Answer:

Section: 2

50. Show how branching could occur during the free-radical polymerization of styrene.

Answer:

Section: 2

51. Provide a mechanism to show how $H_2C=C(CH_3)_2$ is polymerized using BF_3 as the initiatior.

Answer:

Section: 2

52. Provide the structure of Saran, an alternating copolymer of $H_2C=CHCl$ and $H_2C=CCl_2$.

Answer:

Section: 5

MULTIPLE CHOICE

53. What kind of polymer is produced in the following reaction?

    a. a polycarbonate
    b. a polyurethane
    c. a polyester
    d. a synthetic rubber
    e. a poly (acrylonitrile)

$$OCN - \underset{\bigcirc}{} - NCO + HOCH_2CH_2OH \longrightarrow$$

Answer: b Section: 6

SHORT ANSWER

54. Draw the structure of the polymer produced in the following reaction.

Answer:

Section: 6

MULTIPLE CHOICE

55. Natural rubber is a:

    a. polyamide.
    b. polyester.
    c. polycarbonate.
    d. polyurethane.
    e. none of the above.

Answer: e Section: 4

SHORT ANSWER

56. Draw the products which would result if poly(ethylene terephthalate) were hydrolyzed in hot aqueous NaOH.

Answer:

$HOCH_2CH_2OH$ + ^-O_2C—⟨benzene ring⟩—CO_2^-

Section: 6

57. Propose a synthesis of poly(vinyl alcohol) from $H_2C{=}CHO_2CCH_3$.

Answer:
1. $(PhCO_2)_2$
2. $NaOH, H_2O$
Section: 2

58. Draw the structure of the polymer produced in the following reaction.

$\xrightarrow{TiCl_4, (CH_3CH_2)_3Al}$

Answer:

Section: 4

59. Provide a mechanism to show how $H_2C{=}CHCO_2CH_3$ is polymerized using butyllithium as the initiator.

Answer:

Section: 2

631

60. Why is chain branching less common in anionic polymerization than in cationic polymerization?

    Answer:
    The intermediate which leads to chain branching is a more substituted carbon ion. Greater substitution stabilizes carbocations but destabilizes carbanions.
    Section: 2

61. Draw the structure of the polymer and byproduct that result when the two monomers below are heated together.

    $$H_3CO_2C - \langle \bigcirc \rangle - CO_2CH_3 \qquad + \qquad HOCH_2CH_2OH$$

    Answer:

    $$-\left[-OCH_2CH_2O - \overset{O}{\underset{\|}{C}} - \langle \bigcirc \rangle - \overset{O}{\underset{\|}{C}} - \right]_n \qquad + \qquad CH_3OH$$

    Section: 1

MULTIPLE CHOICE

62. From which of the following compounds is Lucite or poly(methyl methacrylate) prepared?

    a. $CH_2=CHCH_3$
    b. $CH_2=C(CH_3)O_2CCH_3$
    c. $CH_2=CHO_2CCH_3$
    d. $CH_2=C(CH_3)CO_2CH_3$
    e. $CH_2=CHCO_2CH_3$

    Answer: d   Section: 2

SHORT ANSWER

63. What is the main structural difference between high-density and low-density polyethylene, and how does this structural difference affect the properties and uses of these polymers?

    Answer:
    High-density polyethylene is linear while low-density polyethylene is branched. Linear, unbranched chains can pack together more closely than branched ones. Thus, high-density polyethylene is a relatively hard plastic and is used for the production of things like artificial hip joints. Low-density polyethylene is a much more flexible polymer which is used to make trash bags and dry cleaning bags.
    Section: 2

64. What advantage does a Lewis acid like BF₃ have over a proton-donating acid like HCl as a initiator in cationic polymerization?

Answer:
The advantage of a Lewis acid initiator like BF_3 is that it does not have an accompanying nucleophile that could act as a chain terminator.

Section: 2

MULTIPLE CHOICE

65. Which of the following can result in chain termination in cationic polymerization?

a. Loss of H+.
b. Addition of a nucleophile that reacts with the propagating site.
c. A chain transfer reaction with the solvent.
d. A 1,2-hydride shift.
e. a, b, and c.

Answer: e   Section: 2

66. Which of the following monomers is least likely to undergo cationic polymerization?

a. propylene
b. isobutylene
c. styrene
d. methyl acrylate
e. vinyl acetate

Answer: d   Section: 2

SHORT ANSWER

67. Is acrylonitrile more likely to undergo anionic or cationic polymerization? Explain your answer.

Answer:
Acrylonitrile is more likely to undergo anionic polymerization. The cyano group is an electron-withdrawing group that can stabilize an anion.

Section: 2

68. Draw the structure of the polymer that results when methyl ?-cyanoacrylate undergoes polymerization. By what name is this polymer commonly known?

    Answer: This polymer is commonly know as Super Glue.

    Section: 2

MULTIPLE CHOICE

69. Which of the following statements concerning the stereochemistry of polymerization is (are) correct?

    a. Polymers in the atactic configuration are more likely to be crystalline solids because regular packing of the polymer chains is easier.
    b. Polymers in the isotactic configuration tend to be softer and more flexible than atactic polymers.
    c. In general, radical polymerization leads primarily to branched polymers in the syndiotactic configuration.
    d. The percentage of chains in the isotactic or syndiotactic configuration decreases as the polymerization temperature decreases.
    e. None of the above.

    Answer: e   Section: 3

70. From which of the following monomers is neoprene synthesized?

    a. (Z)-2-chloro-2-butene
    b. (E)-2-chloro-2-butene
    c. (E)-1-chloro-1,3-butadiene
    d. 2-chloro-1,3-butadiene
    e. 2,3-dichloro-1,3-butadiene

    Answer: d   Section: 4

SHORT ANSWER

71. Draw the structure of Nylon 66.

Answer:

$$\left[-NH(CH_2)_6 NH\overset{\overset{\displaystyle O}{\|}}{C}(CH_2)_4 \overset{\overset{\displaystyle O}{\|}}{C} - \right]_n$$

Section: 6

72. What are thermosetting polymers?

Answer:
Thermosetting polymers cannot be remelted after they have hardened because the cross-links present are covalent bonds and not intermolecular van der Waals forces.

Section: 7

Chapter 27:   Heterocylic Compounds

MULTIPLE CHOICE

1. Which of the following statements best describes the meaning of the term alkaloid?

   a. Heterocyclic natural products with a nitrogen as a heteroatom
   b. Natural, cyclic hydrocarbons
   c. Heterocyclic natural products with a sulfur as a heteroatom
   d. Heterocyclic natural products with an oxygen as a heteroatom
   e. Natural, cyclic hydrocarbons with alkyl groups

   Answer: a Section: 98

2. What is the systematic name for the following compound?

   a. aziridine
   b. oxetane
   c. oxirane
   d. azetidine
   e. thiirane

   Answer: c Section: 1

3. What is the systematic name for the following compound?

   a. aziridine
   b. oxetane
   c. oxirane
   d. azetidine
   e. thiirane

   Answer: e Section: 1

4. Which of the following is the structure for azetidine?

    a. I
    b. II
    c. III
    d. IV
    e. V

I. ☐—NH

II. △ (O)

III. △ (N—H)

IV. ⬠ (N—H)

V. ☐—S

Answer: a   Section: 1

5. Which of the following is the structure for N-methylpiperidine?

   a. I
   b. II
   c. III
   d. IV
   e. V

I.

II.

III.

IV.

V. CH₃—⟨ ⟩—N—H

Answer: b Section: 1

6. Which of the following is the structure of 1,4-dioxane?

    a. I
    b. II
    c. III
    d. IV
    e. V

I.

II.

III.

IV.

V.

Answer: d Section: 1

7. What is the systematic name of the following compound?

    a. 1-Azabicyclo[2.2.2]heptane
    b. 2-Azatricyclo[2.2.2]heptane
    c. 2-Azabicyclo[2.2.2]octane
    d. 2-Azabicyclo[2.2.2]heptane
    e. 2-Azabicyclo[2.2.0]octane

Answer: c Section: 1

8. What is the <u>major</u> product of the following reaction?

   a. I
   b. II
   c. III
   d. IV
   e. V

I. $HNCH_2\overset{\underset{\displaystyle CH_3}{|}}{C}H$ with SCH_3 and CH_3

II. aziridine with CH_3, N–H, SCH_3, CH_3, H

III. aziridine with SCH_3 on N, CH_3, CH_3

IV. $H_3\overset{\oplus}{N}CH\!-\!\overset{\underset{\displaystyle CH_3}{CH_3}}{C}H$ with SCH_3

V. $H_3\overset{\oplus}{N}CH_2\overset{\underset{\displaystyle CH_3}{\overset{\displaystyle CH_3}{|}}}{C}SCH_3$

Answer: e   Section: 1

9. What is the <u>major</u> product of the following reaction?

   a. I
   b. II
   c. III
   d. IV
   e. V

$$CH_3 \underset{CH_3}{\overset{O}{\triangle}} + CH_3S^- \xrightarrow[\text{alcohol}]{\text{heat}} \ ?$$

I. $OCH_2\underset{\underset{SCH_3}{|}}{CH}\overset{CH_3}{\underset{CH_3}{<}}$

II. $CH_3 \underset{CH_3}{\overset{O}{\triangle}} \overset{SCH_3}{\underset{H}{<}}$

III. $CH_3 \underset{CH_3}{\overset{\overset{SCH_3}{|}}{\triangle}}$

IV. $CH_2-\underset{\underset{SCH_3}{|}}{C}\overset{CH_3}{\underset{CH_3}{<}}-OH$

V. $HOCH_2\underset{\underset{CH_3}{|}}{\overset{\overset{CH_3}{|}}{C}}SCH_3$

Answer: d   Section: 1

10. What is the <u>major</u> product of the following reaction?

    a. I
    b. II
    c. III
    d. IV
    e. V

$$\text{(thiirane)} \xrightarrow[\text{2. } H_3O^+]{\text{1. } CH_3MgBr} \; ?$$

I. CH_3—$\underset{CH_3}{\overset{SH}{\underset{|}{C}}}$$CH_2CH_3$

II. CH_3—$\underset{CH_3}{\overset{CH_3}{\underset{|}{C}}}$$CH_2SH$

III. $\underset{CH_3}{\overset{CH_3}{\diagup}}$$CHCH_2SCH_3$

IV. $\overset{CH_3}{\overset{|}{\oplus S}}$ cyclopropane with CH_3, CH_3, CH_3

V. CH_3, CH_3, CH_3 thiirane with S

Answer: a   Section: 1

11. What is the <u>major</u> product of the following reaction?

    a. I
    b. II
    c. III
    d. IV
    e. V

$$\underset{CH_3}{\triangle}\!\!\!^{O} + H_2O^{18} \xrightarrow{\ H^{+}\ } \ ?$$

I. $CH_3CHCH_2\overset{18}{O}H$
          |
         OH

II. $CH_3CHCH_2\overset{18}{O}H$
          |
        $_{18}OH$

III. CH_3CHCH_2OH
          |
        $_{18}OH$

IV. $\underset{CH_3}{\triangle}\!\!\!^{\overset{18}{O}}OH$

V. $\underset{CH_3}{\triangle}\!\!\!^{O}\,_{18}OH$

Answer: c   Section: 1

12. What is the product of the following reaction?

    a. I
    b. II
    c. III
    d. IV
    e. V

$$\text{(lactone)} + H_2^{18}O \xrightarrow{\ H^+\ } \ ?$$

I. $HO^{18}CH_2CH_2CH_2CH_2C\overset{\overset{O}{\|}^{18}}{}OH$

II. $HO^{18}CH_2CH_2CH_2CH_2\overset{\overset{O}{\|}}{C}OH$

III. $HO^{18}CH_2CH_2CH_2CH_2\overset{\overset{^{18}O}{\|}^{18}}{C}OH$

IV. $HOCH_2CH_2CH_2CH_2\overset{\overset{O}{\|}^{18}}{C}OH$

V. $HO^{18}CH_2CH_2CH_2CH_2\overset{\overset{^{18}O}{\|}}{C}OH$

Answer: d   Section: 1

13. What is the <u>major</u> product of the following reaction?

    a. I
    b. II
    c. III
    d. IV
    e. V

I.

II.

III.

IV.

V.

Answer: b Section: 1

14. Which of the following aromatic compounds is the <u>most</u> stable?

    a. I
    b. II
    c. III
    d. IV
    e. V

Answer: d Section: 2

15. What is the <u>major</u> product of the following reaction?

    a. I
    b. II
    c. III
    d. IV
    e. V

Answer: e Section: 2

16. Which of the following compounds is the <u>most</u> reactive toward electrophilic aromatic substitution?

    a. I
    b. II
    c. III
    d. IV
    e. V

    I.

    II.

    III.

    IV.

    V.

    Answer: c Section: 2

17. What is the <u>major</u> product of the following reaction?

 a. I
 b. II
 c. III
 d. IV
 e. V

I.

II.

III.

IV.

V.

Answer: b Section: 2

18. Which of the following compounds is indole?

   a. I
   b. II
   c. III
   d. IV
   e. V

I.

II.

III.

IV.

V.

Answer: e Section: 2

19. What is the <u>major</u> product of the following reaction?

    a. I
    b. II
    c. III
    d. IV
    e. V

Benzofuran

I.

II.

III.

IV.

V.

Answer: a   Section: 2

20. What is the <u>major</u> product of the following reaction?

    a. I
    b. II
    c. III
    d. IV
    e. V

$$+ \; CH_3Br \longrightarrow \; ?$$

I.

II.

III.

IV.

V.

Answer: b Section: 3

21. What is the <u>major</u> product of the following reaction?

    a. I
    b. II
    c. III
    d. IV
    e. V

I.

II.

III.

IV.

V.

Answer: c Section: 3

22. Which of the following is the <u>best</u> method for preparing the following compound?

   a. I
   b. II
   c. III
   d. IV
   e. V

Answer: d Section: 3

23. What are the products of the following reaction?

    a. I and II
    b. I and III
    c. II and IV
    d. II and III
    e. III and IV

quinoline

I.

III.

II.

IV.

Answer: c   Section: 3

24. What is/are the product(s) of the following reaction?

    a. I
    b. II
    c. III
    d. a and b
    e. b and c

I.

II.

III.

Answer: b   Section: 3

25. What is the <u>major</u> product of the following reaction?

   a. I
   b. II
   c. III
   d. IV
   e. V

Answer: a   Section: 4

26. Which of the following is the structure of purine?

a. I
b. II
c. III
d. IV
e. V

I.

II.

III.

IV.

V.

Answer: b Section: 4

27. Which of the following nucleic acids contain the pyrimidine ring structure?

a. adenine, cytosine, and guanine
b. guanine, adenine, and uracil
c. adenine, uracil, and thymine
d. guanine, uracil, and thymine
e. cytosine, uracil, and thymine

Answer: e Section: 4

28. What is the <u>major</u> product of the following reaction?

    a. I
    b. II
    c. III
    d. IV
    e. V

Answer: c Section: 4

29. Which of the following heterocyclics has the lowest pKa value?

   a. I
   b. II
   c. III
   d. IV
   e. V

   I.

   II.

   III. :N⏜:NH

   IV. HN⁺⏜:NH

   V.

   Answer: d   Section: 4

SHORT ANSWER

30. Propose a plausible mechanism for the following reaction.

   + D₂SO₄ ⟶ + HDSO₄

   Answer:

   Section: 2

31. Do you expect pyridine to undergo electrophilic substitution reactions?
    Give an example.

Answer:
Pyridine is very unreactive towards electrophilic substitution
reactions. It is even less reactive than nitrobenzene. However, under
rigorous conditions, substitution on the C-3 position will occur with
low yield.

Section: 3

32. Do you expect pyridine to undergo nucleophilic substitution reactions?
    If yes, on what position would the substitution occur? Give an example.

Answer:
Pyridine is more reactive than benzene towards nucleophilic substitution
reactions. The substitution takes place on the C-2 and C-4 positions.

Section: 3

33. Imidazole boils at 257° C, while the boiling point of N-methylimidazole is 199° C. Explain the difference in their boiling points.

Imidazole N-methylimidazole

Answer:
Imidazole exhibits hydrogen bonding that N-methylimidazole does not. Therefore, imidazole will boil at a higher temperature than N-methylimidazole.

Section: 4

34. Which of the following compounds is easier to decarboxylate?

Answer:
Compound II would be easier to decarboxylate because the electrons left behind when CO_2 is removed can be delocalized.

Section: 3

35. Arrange the following in decreasing order of reactivity towards electrophilic aromatic substitution.

Answer: B > C > A Section: 98

36. Draw the structure of piperidine.

Answer:

Section: 1

37. Provide the systematic name for the compound below.

Answer: pyrrole Section: 2

38. Circle the stronger base in the pair below, and briefly explain your choice.

Answer:
Pyrrolidine is the stronger base. Protonation of pyrrole results in loss of aromaticity.

Section: 2

39. Circle the stronger base in the pair below, and briefly explain your choice.

or $(CH_3CH_2)_2NH$

Answer:
Diethylamine is the stronger base. The lone pair electrons in pyridine are in an sp^2 orbital which is lower in energy than the sp^3 orbital of the lone pair in diethylamine.
Section: 3

40. Circle the stronger base in the pair below, and briefly explain your choice.

NH$_2$ or   NH$_2$

Answer:
Cyclohexylamine is the stronger base. The lone pair electrons in aniline are delocalized into the adjacent aromatic π system.

41. Provide the structure of the major organic product in the reaction below.

Answer:

Section: 2

42. Pyridine does not undergo Friedel-Crafts reactions. Offer an explanation.

Answer:
The presence of the electronegative nitrogen atom in the ring deactivates it toward reaction with electrophiles. In fact, the nitrogen lone pair tends to react with electrophiles to produce an even further deactivated ring. It should also be noted that the nitrogen lone pair is in an orbital perpendicular to the ring's π system and could not stabilize the intermediate carbocation in any event.

Section: 3

MULTIPLE CHOICE

43. When pyrrole undergoes electrophilic aromatic substitution, at which position does substitution occur?

a. 2-position
b. 3-position
c. 4-position

Answer: a Section: 2

44. Which of the following chloropyridines readily undergo nucleophilic substitution upon treatment with NaCN?

    a. 2-chloropyridine
    b. 3-chloropyridine
    c. 4-chloropyridine
    d. both a and b
    e. both a and c
    f. both b and c

    Answer: e   Section: 3

SHORT ANSWER

45. Provide a structural representation of oxetane.

    Answer:

    Section: 1

46. Provide a structural representation of cis-3-ethyl-1,2-epoxycyclopentane.

    Answer:

    CH₂CH₃

    Section: 1

47. Provide the major organic product in the reaction below.

    1. Cl₂, H₂O
    2. NaOH

    Answer:

    Section: 98

48. Provide the major organic product in the reaction below.

$HOCH_2(CH_2)_2CH_2Br \xrightarrow{NaOH}$

Answer:

Section: 98

49. Provide the major organic product in the reaction below.

$$CH_3 \quad CH_3$$
$$H \quad O \quad CH_3 \xrightarrow{CH_3OH, H^+}$$

Answer:

OH
CH₃ — OCH₃
CH₃ — CH₃

Section: 1

MULTIPLE CHOICE

50. What results when cis-2-butene is subjected to the following reaction sequence: (1) Cl_2, H_2O, (2) NaOH, (3) H_3O^+?

a. a meso epoxide
b. a 1:1 mixture of enantiomeric epoxides
c. a meso diol
d. a 1:1 mixture of enantiomeric diols
e. 2-butanol

Answer: d Section: 1

SHORT ANSWER

51. What is the stereochemistry of the product of the acid hydrolysis of trans-2,3-epoxybutane?

Answer: meso diol Section: 1

52. Provide an acceptable name for the compound shown below.

Answer: 1,4-dioxane Section: 1

MULTIPLE CHOICE

53. Which of the following compounds is <u>not</u> aromatic?

    a. I
    b. II
    c. III
    d. IV

I.

II.

III.

IV.

Answer: d Section: 2

SHORT ANSWER

54. Indicate whether the following molecule is aromatic, antiaromatic, or nonaromatic.

Answer: aromatic Section: 2

55. Indicate whether the following molecule is aromatic, antiaromatic, or nonaromatic.

Answer: aromatic Section: 3

56. Indicate whether the following molecule is aromatic, antiaromatic, or nonaromatic.

Answer: aromatic Section: 3

MULTIPLE CHOICE

57. An unknown compound has the formula $C_8H_{10}OS$, and is known to contain a thiophene ring. Thr proton NMR spectrum of this compound is:
    δ0.98, triplet, 3H
    δ1.74, multiplet, 2H
    δ2.80, triplet, 2H
    δ7.04, multiplet, 1H
    δ7.55, multiplet, 2H

    What is the structure of this unknown?

    a. I
    b. II
    c. III
    d. IV

I. $CH_3CH_2CH_2$ —⟨thiophene⟩— $\overset{\overset{O}{\|}}{C}$-H

II. CH_3 —⟨thiophene⟩— $CH_2CH_2\overset{\overset{O}{\|}}{C}$-H

III. ⟨thiophene⟩— $CH_2CH_2\overset{\overset{O}{\|}}{C}$-$CH_3$

IV. ⟨thiophene⟩— $\overset{\overset{O}{\|}}{C}$-$CH_2CH_2CH_3$

Answer: d Section: 98

667

58. What is the major product of the reaction sequence shown?

$$CH_3CHCH_3 \xrightarrow{\text{Mg, ether}} \xrightarrow{\underset{H_2C --- CH_2}{\overset{O}{\diagup \diagdown}}} \xrightarrow{H_3O^+} ?$$
    |
    Cl

a. $(CH_3)_2CHCH_2CH_2OH$
b. $(CH_3)_2CHCHOHCH_3$
c. $CH_3CH_2CH_2CH_2CH_2OH$
d. $CH_3CH_2CH_2CHOHCH_3$

Answer: a   Section: 98

59. In the following question choose the best answer.
The tertiary amine

a. I
b. II
c. III
d. IV

I. N—CH3

II. NHCH3

III. $(CH_3)_3NHCH_3$

IV. $(CH_3)_4NBr$

Answer: a   Section: 1

60. In the following question choose the best answer.
The most basic amine.

    a. I
    b. II
    c. III
    d. IV

    I. $CH_3CH_2NH_2$

    II. $(CH_3)_3N$

    III.

    IV.

Answer: a   Section: 2

SHORT ANSWER

61. Draw the structure of the aziridinium ion and explain why its pKa is lower than a typical secondary ammonium ion.

    Answer:
    In the aziridinium ion, the C-N-C internal bond angle is somewhat smaller than usual which causes the external bond angles to be somewhat larger than usual. The larger external bond angles cause the orbital that the nitrogen uses to overlap with the orbital of hydrogen to have more s character than a typical sp^3 orbital. This increased s character makes this nitrogen more electronegative than the typical secondary ammonium nitrogen.

Section: 1

62. Draw the structures of piperidine and morpholine and explain their relative basicities.

Answer:
Piperidine is a stronger base than morpholine since the presence of the electronegative oxygen in morpholine serves to withdraw electron density from the nitrogen atom.

piperidine   morpholine

Section: 1

MULTIPLE CHOICE

63. The major organic product of the reaction between the compounds below
_____.

a. is an N-acyl pyrrolidine
b. results from an electrophilic aromatic substitution reaction
c. results from a nucleophilic aromatic substitution reaction
d. is a m-disubstituted benzene derivative
e. both b and d

Answer: a   Section: 1

SHORT ANSWER

64. Show how the dipole moments are oriented in pyrrolidine and pyrrole, and explain why they are oriented in opposite directions.

Answer:
The negative end of the molecular dipole in pyrrole is oriented opposite the nitrogen atom. This indicates that nitrogen's ability to donate electrons into the ring by resonance more than makes up for its inductive electron withdrawal.

Section: 2

MULTIPLE CHOICE

65. Which of the following lists the correct order of reactivity of the substrates in electrophilic aromatic substitution reactions?

a. thiophene > pyrrole > furan > benzene
b. benzene > furan > thiophene > pyrrole
c. furan > pyrrole > benzene > thiophene
d. benzene > pyrrole > thiophene > furan
e. pyrrole > furan > thiophene > benzene

Answer: d   Section: 2

66. When benzofuran is treated with acetic anhydride, the major organic product is _____.

a. 2-acetylbenzofuran
b. 3-acetylbenzofuran
c. 4-acetylbenzofuran
d. 5-acetylbenzofuran
e. 7-acetylbenzofuran

Answer: a   Section: 2

67. Which of the following statements correctly describes the behavior of heterocycles in aromatic substitution reactions?

   a. Pyridine N-oxide is considerably more reactive toward electrophilic aromatic substitution reactions than pyridine because the oxygen atom can donate electrons into the ring by resonance.
   b. Pyridine N-oxide undergoes electrophilic aromatic substitution primarily at C-3.
   c. Pyridine is less reactive than benzene toward nucleophilic aromatic substitution.
   d. Nucleophilic aromatic substitution of pyridine takes place at C-3 because attack at this position leads to the most stable intermediate.
   e. None of the above.

   Answer: a   Section: 3

SHORT ANSWER

68. Draw the structures of the keto forms of both 2-hydroxypyridine and 4-hydroxypyridine.

   Answer:

   Section: 3

MULTIPLE CHOICE

69. What is the major organic product of the following reaction?

   a. 4-ethyl-2,3-dimethylpyridine
   b. 5-ethyl-2,3-dimethylpyridine
   c. 6-ethyl-2,3-dimethylpyridine
   d. 2-methyl-3-propylpyridine
   e. 3-methyl-2-propylpyridine

   Answer: e   Section: 3

SHORT ANSWER

70. Draw the structure of a porphyrin ring system, and list three common, naturally-occurring molecules which contain a porphyrin-like ring system as part of their structure.

Answer: heme; chlorophyll a; vitamin B_{12}

Section: 4

Chapter 28:   Pericyclic Reactions

MULTIPLE CHOICE

1. Which of the following statements best describes what is meant by pericyclic reactions?

   a. Reactions in which a nucleophile reacts with an electrophile
   b. Reactions in which an electrophile reacts with a nucleophile
   c. Reactions in which a free radical reacts with another free radical
   d. Oxidation-reduction reactions in which electrons are lost or gained
   e. Reactions that occur as a result of a cyclic reorganization of electrons

   Answer: e Section: 98

2. Which of the following reactions best describes the Diels-Alder reaction?

   a. Electrocyclic reaction
   b. Cycloaddition reaction
   c. Sigmatropic reaction
   d. Radical reaction
   e. Nucleophilic substitution reaction

   Answer: b Section: 4

3. Which of the following statements best describes the theory of Conservation of Orbital Symmetry?

   a. Molecular orbital of the transition state must be similar to that of the reactant.
   b. Molecular orbital of the transition state must be similar to that of the product.
   c. Only s orbitals from reactants and products are utilized.
   d. Molecular orbitals of reactant and product must have similar symmetry.
   e. Molecular orbitals of reactant and product must have different symmetry.

   Answer: d Section: 1

4. Which of the following pericyclic reactions best describes the following reaction?

   a. Nucleophilic
   b. Electrocyclic
   c. Radical
   d. Sigmatropic
   e. Cycloaddition

Answer: e Section: 4

5. Which of the following molecular orbitals is produced when in-phase atomic orbitals overlap?

   a. Bonding molecular orbital
   b. Antibonding molecular orbital
   c. Nonbonding molecular orbital
   d. Out-of-phase molecular orbital
   e. Excited energy state of molecular orbital

Answer: a Section: 2

6. Which of the following molecular orbitals are formed from <u>subtractive</u> overlap of atomic orbitals?

   a. Antibonding orbitals
   b. Out-of-phase orbitals
   c. Bonding orbitals
   d. a and b
   e. b and c

Answer: d Section: 2

7. How many molecular orbitals are produced from the overlap of four p atomic orbitals?

   a. 2
   b. 8
   c. 4
   d. 16
   e. 1

Answer: c Section: 2

8. Which of the following sets of atomic orbitals form an asymmetric molecular orbital?

   a. I
   b. II
   c. III
   d. IV
   e. V

Answer: e Section: 2

9. How many π molecular orbitals does 1,3,5,7-octatetraene have?

   a. 2
   b. 4
   c. 6
   d. 8
   e. 10

Answer: d Section: 2

10. How many nodes does the highest energy molecular orbital of 1,3,5,7-octatetraene have?

   a. 4
   b. 5
   c. 6
   d. 7
   e. 8

Answer: e Section: 2

11. What is the <u>major</u> product of the following reaction?

   a. I
   b. II
   c. III
   d. IV
   e. V

I.

II.

III.

IV.

V.

Answer: b Section: 3

12. What is the major product of the following reaction?

    a. I
    b. II
    c. III
    d. IV
    e. V

    I.

    II.

    III.

    IV.

    V.

    Answer: b   Section: 3

13. Which of the following statements best describes the term <u>suprafacial</u>?

    a. Formation of two sigma bonds from the same side of the pi system
    b. Formation of two pi bonds from the same side of the sigma system
    c. Formation of two sigma bonds from opposite sides of the pi system
    d. Formation of two pi bonds from opposite sides of the sigma system
    e. An anti cycloaddition

    Answer: a   Section: 4

14. Which of the following ring formations must involve a suprafacial bond formation?

    a. four-membered ring
    b. five-membered ring
    c. six-membered ring
    d. a and c
    e. a, b and c

    Answer: e   Section: 4

15. What is the <u>major</u> product of the following reaction?

    a. I
    b. II
    c. III
    d. IV
    e. V

Answer: c Section: 4

16. Which of the following best describes the sigmatropic rearrangement that occurs in the reaction shown below?

    a. [1,3] Sigmatropic rearrangement
    b. [2,3] Sigmatropic rearrangement
    c. [3,3] Sigmatropic rearrangement
    d. [1,5] Sigmatropic rearrangement
    e. [1,4] Sigmatropic rearrangement

Answer: d Section: 5

17. What is the major product of the following [3,3] sigmatropic rearrangement?

    a. I
    b. II
    c. III
    d. IV
    e. V

Answer: e   Section: 5

18. What is the major product from the following rearrangement?

    a. I
    b. II
    c. III
    d. IV
    e. V

I.

II.

III.

IV.

V.

Answer: a   Section: 5

19. Under what conditions would a 1,3-hydrogen shift occur in a sigmatropic rearrangement?

    a. Thermal conditions
    b. Photochemical conditions
    c. Suprafacial pathway
    d. a and c
    e. b and c

Answer: e   Section: 5

20. What is/are the product(s) of the following 1,3-hydrogen shift?

    a. I
    b. II
    c. III
    d. I and II
    e. I and III

Answer: d   Section: 5

21. Under what conditions would a 1,5-hydrogen shift occur in a sigmatropic rearrangement?

    a. Thermal conditions
    b. Photochemical conditions
    c. Suprafacial pathway
    d. a and c
    e. b and c

Answer: d   Section: 5

22. What is the product of the following 1,5-hydrogen shift?

$$CH_3CH=CHCH=CD_2 \xrightarrow{\Delta} ?$$

    a. $CH_2=CHCH_2CH=CHD$
    b. $CH_2=CHCH=CHCHD_2$
    c. $CHD=CHCH=CHCH_3$
    d. $CH_2=CHCH_2CH_2CH_2D$
    e. $CH_2=CHCH=CHCD_3$

Answer: b   Section: 5

23. What is the product of the following 1,7-hydrogen shift?

   a. I
   b. II
   c. III
   d. IV
   e. V

I.

II.

III.

IV.

V.

Answer: c Section: 5

24. What is the product from the following 1,3-carbon shift reaction?

    a. I
    b. II
    c. III
    d. IV
    e. V

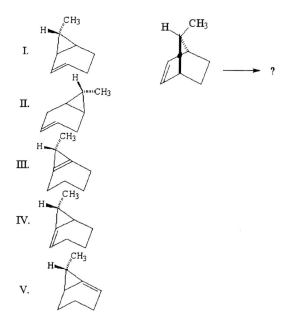

Answer: a Section: 5

25. What is the product obtained from the following [1,5] sigmatropic rearrangement?

    a. I
    b. II
    c. III
    d. IV
    e. V

Answer: d Section: 5

26. Which of the following best explains the rearrangement when provitamin D_3 rearranges to vitamin D_3?

    a. [1,7] cycloaddition rearrangement, antarafacial
    b. [1,7] cycloaddition rearrangement, suprafacial
    c. [1,7] sigmatropic rearrangement, antarafacial
    d. [1,7] sigmatropic rearrangement, suprafacial
    e. [1,7] electrocyclic reaction, suprafacial

Provitamin D_3 Vitamin D_3

Answer: c Section: 5

SHORT ANSWER

27. What are the three most common pericyclic reactions? Give examples.

  Answer:

  a. electrocyclic reactions:

  b. cycloaddition reactions:

  c. sigmatropic rearrangement:

  Section: 1

28. Show the mechanism for each of the following pericyclic reactions.

  Answer:

  Section: 1

29. What is meant by <u>frontier orbitals</u>?

    Answer:
    Frontier orbitals are the two molecular orbitals known as the highest occupied molecular orbital (HOMO) and the lowest unoccupied molecular orbital (LUMO).

    Section: 2

30. Indicate whether each of the following reactions is an electrocyclic reaction, a cycloaddition reaction, or a sigmatropic rearrangement.

    a.

    b.

    c.

    Answer:
    a. Is an electrocyclic reaction.
    b. Is a sigmatropic rearrangement.
    c. Is a cycloaddition reaction.

    Section: 1

31. What does LCAO stand for?

    Answer: Linear combination of atomic orbitals. Section: 2

32. Which molecular orbital is higher in energy: the in-phase or out-of-phase overlap of atomic orbitals?

    Answer:
    Out-of-phase overlap of atomic orbitals produces an antibonding molecular orbital which is higher in energy than that of the bonding molecular orbital.

    Section: 2

33. Explain what a node is and how it is formed.

Answer:
A node is a plane in which there is no probability of finding an electron. It is formed between overlapping out-of-phase orbitals or through the center of p orbitals.

Section: 2

34. How many nodes would you expect from the following atomic orbital overlap?

Answer:
The molecular orbital is an out-of-phase overlap producing four nodes (including the molecule's nodal plane).

Section: 2

35. (a) Under thermal conditions, will ring closure of (2E, 4Z, 6Z, 8E)-decatetraene be conrotatory or disrotatory? Explain.
    (b) Will the product be cis or trans? Explain

Answer:
(a) Since the number of conjugated pi bonds is even and the reaction occurs under thermal conditions, the Woodward-Hoffman rules for electrocyclic reaction predicts that the ring closure is conrotatory.

(b) Since the ring closure is conrotatory and the closure is (E,E) configuration, the product will have a trans configuration.

Section: 3

36. Explain why the following (2+2) cycloaddition reaction would work under photochemical conditions but not under thermal conditions.

Answer:
Under thermal conditions, suprafacial overlap is not symmetry-allowed while under photochemical conditions, the reaction can take place because it involves symmetry-allowed suprafacial bond formation.
Section: 4

37. Show the mechanism for the following 1,3-hydrogen shift.

Answer:

Section: 5

MULTIPLE CHOICE

38. In the allyl radical, which π molecular orbital is singly occupied?

   a. The bonding π molecular orbital.
   b. The nonbonding π molecular orbital.
   c. The antibonding π molecular orbital.
   d. None of the above.

Answer: b Section: 2

SHORT ANSWER

39. Give a representation of the bonding π molecular orbital of the allyl cation.

Answer:

Section: 2

40. Give a representation of the antibonding π molecular orbital of the allyl anion.

Answer:

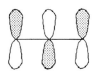

Section: 2

MULTIPLE CHOICE

41. How many electrons are present in the nonbonding π molecular orbital of the allyl anion?

    a. 0
    b. 1
    c. 2
    d. 3
    e. 4

    Answer: c Section: 2

SHORT ANSWER

42. What diene and dienophile would react to give the product below?

Answer:

Section: 4

43. Provide the structure of the major organic product in the following reaction.

Answer:

Section: 4

44. What are the structures of A, B, and C?

A $\xrightarrow{\text{NBS}}$ A $\xrightarrow[\text{CH}_3\text{CH}_2\text{OH}, \Delta]{\text{NaOCH}_2\text{CH}_3}$ B

$\xrightarrow{\text{CH}_2=\text{CHCO}_2\text{H}}$ C $\xrightarrow[\text{2. S(CH}_3)_2]{\text{1. O}_3}$

Answer:

A   B   C

Section: 4

45. Why does the diene shown below fail to undergo a Diels-Alder reaction with even the most reactive dienophiles?

Answer: The diene cannot achieve the necessary s-cis conformation.

Section: 4

46. What characterizes a pericyclic reaction?

Answer:
Pericyclic reactions involve concerted breaking and formation of bonds within a closed ring of interacting orbitals.

Section: 98

47. Consider the possible thermal [4+4] cycloaddition of two molecules of 1,3-butadiene to generate cycloocta-1,5-diene. Show the HOMO/LUMO interaction which would result, and use this interaction to predict whether the proposed cycloaddition could occur.

Answer:

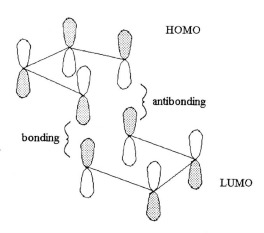

The cycloaddition is thermally forbidden.

Section: 4

48. Is the thermal [4+2] cycloaddition between allylanion and ethylene an allowed one? To answer, draw the HOMO of allyl anion and the LUMO of ethylene, and comment on the symmetry match of the two.

Answer:

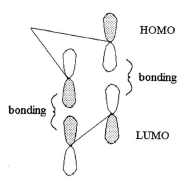

The cycloaddition is thermally allowed.

Section: 4

MULTIPLE CHOICE

49. Through what type of single pericyclic reaction did the reaction shown below proceed?

    a. [1,2] sigmatropic hydrogen migration
    b. [1,3] sigmatropic hydrogen migration
    c. [1,4] sigmatropic hydrogen migration
    d. [1,5] sigmatropic hydrogen migration
    e. None of the above

Answer: d Section: 1

SHORT ANSWER

50. Consider the π molecular orbitals of pentadienyl cation and answer the following questions. Please refer to the lowest π MO as Ψ_1.
    a. How many π molecular orbitals of pentadienyl cation are bonding?
    b. Which π molecular orbital is the HOMO?
    c. Which π molecular orbitals are asymmetric with respect to a perpendicular mirror plane?

    Answer:
    a Ψ_1 and Ψ_2.
    b. Ψ_2.
    c. Ψ_2 and Ψ_4.
    Section: 2

51. Electrocyclic reactions are reversible. The cyclic compound is favored for electrocyclic reactions that form six-membered rings, whereas the open-chain compound is favored for reactions that form four-membered rings. Explain thoroughly in terms of the enthalpy changes that are occurring.

    Answer:
    In general, the cyclic compound is favored on enthalpy grounds because sigma bonds are lower in energy than pi. However, in the case of four-membered ring formation this favorable enthalpy change is offset by an unfavorable enthalpy change as significant angle and torsional strain are introduced in the four-membered ring.

    Section: 3

MULTIPLE CHOICE

52. Which of the following statements concerning electrocyclic reactions is (are) correct?

   a. The ground state HOMO of a compound with an even number of conjugated double bonds is asymmetric.
   b. The allowed mode of ring closure under thermal conditions for a molecule containing an even number of conjugated double bonds is disrotatory.
   c. The allowed mode of ring closure under photochemical conditions for a molecule containing an odd number of conjugated double bonds is disrotatory.
   d. Both b and c.
   e. None of the above.

   Answer: a   Section: 3

SHORT ANSWER

53. Draw the structure of the major organic product of the following reaction. Is the product chiral or achiral?

   Answer:

   The product is achiral and meso.

   Section: 3

54. What type of pericyclic process is occurring in the reaction below? Provide two reasons why the equilibrium lies in the direction indicated.

Answer:
A [3,3] sigmatropic rearrangement is occurring. This particular reaction is also known as a Cope rearrangement. The equilibrium lies in the direction indicated because substituting a double bond and conjugating it with an aromatic ring are both stabilizing factors.

Section: 5

MULTIPLE CHOICE

55. Which of the following statements about carbon migrations is (are) correct?

a. Carbon migrations never occur by a sigmatropic rearrangement.
b. Under thermal conditions, a [1,3] carbon migration that occurs suprafacially must also occur with inversion of configuration at the migrating carbon.
c. Under thermal conditions, a [1,5] carbon migration that occurs suprafacially must also occur with inversion of configuration at the migrating carbon.
d. Under thermal conditions, a carbon migration that occur suprafacially must occur with inversion no matter the system's symmetry.
e. b, c, and d.

Answer: b   Section: 5

SHORT ANSWER

56. What pericyclic photochemical reaction can lead to skin cancer in humans? Draw the structure of the compounds involved in the reaction, and explain how this reaction leads to skin cancer.

Answer:
Thymine dimerization

At any point in DNA where there are two adjacent thymine residues, a [2+2] photocycloaddition can occur. This reaction, caused by absorption of a photon of UV light, results in the formation of a thymine dimer. Such structural changes can lead to mutations and possible cancer.

Section: 6

MULTIPLE CHOICE

57. Which of the following statements concerning the selection rules for pericyclic reactions is (are) correct?

a. If TE describes the reaction, then the outcome is given by AC.
b. If PO describes the reaction, then the outcome is given by AC.
c. If TO describes the reaction, then the outcome is given by AC.
d. Both a and b.
e. None of the above.

Answer: d   Section: 1

58. Which of the following statements about the product of the electrocyclic ring closure below is (are) correct?

   a. The methyl groups are *trans* in the product.
   b. The product arose from a stereospecific, conrotatory ring closure.
   c. The product mixture is optically active.
   d. Both b and c.
   e. All of the above.

   Answer: b   Section: 3

Chapter 29: More About Multistep Organic Synthesis

MULTIPLE CHOICE

1. Which of the following statements best explains the meaning of <u>natural products</u>?

    a. Organic compounds that are used in organic gardening
    b. Organic compounds that are harmless to nature
    c. Organic compounds synthesized in nature
    d. Organic compounds synthesized in the lab that resemble those found in nature
    e. Organic compounds synthesized in the lab that degrade safely in nature

    Answer: c Section: 98

2. Why do chemists synthesize organic compounds?

    a. To study their properties
    b. To answer a variety of chemical questions
    c. To study their unusual shape or structure
    d. To provide us with a greater supply of a compound than nature can produce
    e. All of the above

    Answer: e Section: 98

3. Which of the following reasons would make one synthetic method superior to others?

    a. Produces a higher yield
    b. Utilizes fewer steps
    c. Uses less complicated reactions
    d. Requires cheaper starting materials
    e. All of the above

    Answer: e Section: 98

4. Which of the following best describes convergent synthesis?

    a. A → B
    b. A → B → C
    c. A → B → C + D → F
                        ↑
                        E

    d. A → B → C → D → E
    e. B → A

    Answer: c Section: 98

5. Which of the following methods may be used for the synthesis of propyl alcohol,

$$CH_3CH_2CH_2OH ?$$

a.

$$CH_3CH_2\overset{\overset{\displaystyle O}{\|}}{C}H \xrightarrow[\text{2. } H_2O]{\text{1. } NaBH_4}$$

b.

$$CH_3CH_2\overset{\overset{\displaystyle O}{\|}}{C}OH \xrightarrow[\text{2. } H_2O]{\text{1. } LiAlH_4}$$

c.

$$CH_3CH = CH_2 \xrightarrow[\text{2. } HO^-, H_2O_2, H_2O]{\text{1. } BH_3}$$

d.

$$CH_3CH_2CH_2Br \xrightarrow{\,^-OH\,}$$

e. All of the above

Answer: e   Section: 1

6. Which of the following reduction reactions may be used to synthesize benzyl alcohol,

a. I
b. II
c. III
d. I and II
e. I and III

I.

II.

III.

Answer: d   Section: 1

7. Which of the following methods cannot be used to prepare isobutyl alcohol,

$$CH_3CHCH_2OH$$ with a CH_3 group attached to the middle carbon ?

a.

$CH_3C = CH_2$ with CH_3 substituent $\xrightarrow[H^+]{H_2O}$

b.

$CH_3C = CH_2$ with CH_3 substituent $\xrightarrow[2.\ HO^-,\ H_2O_2,\ H_2O]{1.\ BH_3}$

c.

CH_3CHCH_2Br with CH_3 substituent $\xrightarrow{HO^-}$

d.

CH_3CHCH with CH_3 substituent and O double bond $\xrightarrow[2.\ H_2O]{1.\ NaBH_4}$

e.

$CH_3CHCH_2OCH_3$ with CH_3 substituent $\xrightarrow[heat]{HI}$

Answer: a   Section: 1

8. Which of the following statements best explains the meaning of retrosynthesis?

a. Working forward from the starting material to the desired product
b. Working backward from the desired product to the starting material
c. Working forward using linear synthesis
d. Working forward using convergent synthesis
e. Dissecting the starting material into smaller and smaller pieces until the desired product is reached

Answer: b   Section: 2

9. Which of the following statements best explains what is meant by
   synthons?

   a. Fragments of a connection in a convergent synthesis
   b. Fragments of a connection in a linear synthesis
   c. Fragments of a disconnection in retrosynthesis
   d. Bond formation in a retrosynthesis
   e. Bond formation in a linear synthesis

   Answer: c Section: 2

10. Which of the following statements best explains what is meant by a
    synthetic equivalent?

    a. Using equal equivalents of reactants
    b. Using equal equivalents of reactants and catalysts
    c. Producing equivalents of the products equal to the equivalents of
       reactants
    d. Actual reagent used as a source of the synthon
    e. Using a microscale approach to retrosynthesis

    Answer: d Section: 2

11. Which of the following might be considered synthons for the retrosynthetic analysis of the compound shown below?

    a. I
    b. II
    c. III
    d. IV
    e. All of the above

Answer: e Section: 2

12. Which of the following is the best synthetic equivalent of $:H^-$?

    a. H_2O
    b. $LiAlH_4$
    c. HO^-
    d. $:^-NH_2$
    e. PH_3

Answer: b Section: 2

13. Which of the following statements best explains the meaning of <u>umpolung</u>?

    a. Retention of configuration in a functional group
    b. Interconversion of a functional group
    c. Retaining the normal polarity of a functional group
    d. Reversing the normal polarity of a functional group
    e. Enhancing the normal polarity of a functional group

Answer: d Section: 2

14. What is the synthetic equivalent for the following synthon?

$CH_3CH_2{:}^{\ominus}$

  a. CH_3CH_2Br
  b.
    $CH_3CH_2\overset{..}{N}HCH_2CH_3$
  c. CH_3CH_2MgBr
  d. CH_3CH_2ONa
  e.
    $CH_3CH_2\overset{..}{\underset{..}{O}}CH_2CH_3$

Answer: c Section: 2

15. What is the synthetic equivalent for

  a. I
  b. II
  c. III
  d. IV
  e. V

$$^{\ominus}{:}CH_2\overset{O}{\overset{\|}{C}}H \quad ?$$

I. $CH_3\overset{O}{\overset{\|}{C}}H$

II. $\underset{Cl}{CH_2}\overset{O}{\overset{\|}{C}}H$

III. $\underset{CH_3}{CH_2}\overset{O}{\overset{\|}{C}}H$

IV. (cyclohexene)$CH_2\overset{O}{\overset{\|}{C}}H$

V. (benzene)$CH_2\overset{O}{\overset{\|}{C}}H$

Answer: a Section: 2

16. How would you accomplish the following synthesis?

    a. I
    b. II
    c. III
    d. IV
    e. V

I.

II.

III.

IV.

V.

Answer: e Section: 2

17. Which of the following is not commonly used as a reducing agent?

    a. H_2
    b. H^-
    c. H^+
    d. Zn
    e. Fe

Answer: c Section: 2

18. How can you synthesize the following compound from starting materials containing <u>no</u> <u>more</u> <u>than</u> <u>four</u> carbons?

    a. I
    b. II
    c. III
    d. IV
    e. V

Answer: b Section: 1

19. Propose a method to synthesize the following compound:

    a. I
    b. II
    c. III

I.

II.

III.

Answer: e Section: 2

20. Which of the following statements explains the meaning of a <u>protecting group</u>?

    a. A reagent that interconverts a functional group during a synthetic operation
    b. A reagent that increases the reactivity of a functional group during a synthetic operation
    c. A reagent that protects a functional group from a synthetic operation
    d. A reagent that decreases the reactivity of a functional group during a synthetic conversion
    e. A reagent that improves the availability of a functional group for synthetic conversions

Answer: c Section: 4

21. Identify the protecting group in the following synthesis:

a. I
b. II
c. III
d. IV
e. V

I.

II.   H₃O⁺

III.

IV.   HOCH₂CH₂CH₂OH
V.   LiAlH₄/H₂O

Answer: d Section: 4

22. Which of the following protecting groups is best used to protect the carboxyl group in the following synthesis?

a. I
b. II
c. III
d. IV
e. V

I. $HSCH_2CH_2SH$

II. $HOCH_2CH_2CH_2OH$

III. CH_3CH_2OH

IV.

$$CH_3-\overset{\overset{\displaystyle NH_2}{|}}{\underset{\underset{\displaystyle CH_3}{|}}{C}}CH_2OH$$

V. $CH_3\overset{\overset{\displaystyle O}{||}}{C}Cl$

Answer: d Section: 4

23. Which of the following protecting groups is best used to protect the amino group in the following reaction?

    a. I
    b. II
    c. III
    d. IV
    e. V

I. $HSCH_2CH_2SH$

II. $HOCH_2CH_2CH_2OH$

III. CH_3CH_2OH

IV.
$$CH_3-\overset{\overset{\displaystyle NH_2}{|}}{\underset{\underset{\displaystyle CH_3}{|}}{C}}CH_2OH$$

V.
$$CH_3\overset{\overset{\displaystyle O}{||}}{C}Cl$$

Answer: e   Section: 4

24. Which of the following statements best explains the meaning of an <u>enantioselective reaction</u>?

    a. A reaction that forms the S configuration exclusively
    b. A reaction that forms a racemic mixture
    c. A reaction that forms diastereomers
    d. A reaction that forms optically inactive products
    e. A reaction that forms an excess of one enantiomer over the other

Answer: e   Section: 5

25. How would the following synthesis best be accomplished?

$$CH_3CH_2C\equiv CH \longrightarrow \underset{\text{trans-3-hexene}}{CH_3CH_2,H \atop C=C \atop H,CH_2CH_3}$$

1-Butyne

a. I
b. II
c. III
d. IV
e. V

$$CH_3CH_2C\equiv CH \longrightarrow \underset{\text{trans-3-hexene}}{CH_3CH_2 \atop C=C \atop H, CH_2CH_3}$$
1-Butyne

I. $CH_3CH_2C\equiv CH + Na \longrightarrow CH_3CH_2C\equiv \overset{\ominus}{\underset{\oplus}{C}Na} \xrightarrow{CH_3CH_2Br} CH_3CH_2C\equiv CCH_2CH_3$
    $\xrightarrow[\substack{\text{Lindlar's} \\ \text{catalyst}}]{H_2}$ Product

II. $CH_3CH_2C\equiv CH + Na \longrightarrow CH_3CH_2C\equiv \overset{\ominus}{\underset{\oplus}{C}Na} \xrightarrow{CH_3CH_2Br} CH_3CH_2C\equiv CCH_2CH_3$
    $\xrightarrow[NH_3]{Li}$ Product

III. $CH_3CH_2C\equiv CH + CH_3CH_2MgBr \longrightarrow CH_3CH_2C\equiv CCH_2CH_3$
    $\xrightarrow[\substack{\text{Lindlar's} \\ \text{catalyst}}]{H_2}$ Product

IV. $CH_3CH_2C\equiv CH + CH_3CH_2MgBr \longrightarrow CH_3CH_2C\equiv CCH_2CH_3$
    $\xrightarrow[NH_3]{Li}$ Product

V. $CH_3CH_2C\equiv CH + CH_3CH_2 \xrightarrow{Na} CH_3CH_2C\equiv CCH_2CH_3$
    $\xrightarrow[NH_3]{Na}$ Product

Answer: b Section: 1

26. Identify the <u>chiral</u> <u>auxiliary</u> in the following synthesis:

    a. I
    b. II
    c. III
    d. IV
    e. V

I.

II. LDA

III. CH₃CH₂I

IV.

V. H₃O⁺

Answer: a   Section: 5

27. Which of the following methods may be used to produce an <u>enantioselective</u> reaction?

    a. Using an enzyme-catalyzed reaction
    b. Using a chiral auxiliary
    c. Using enantiomerically pure starting material
    d. Using a reactant that already has the chirality center with the configuration needed in the product
    e. All of the above

Answer: e   Section: 5

28. What is meant by an <u>iterative</u> synthesis?

    a. A reaction in which the product can be easily reversed to the reactant
    b. A synthesis in which a catalyst is never used
    c. A synthesis in which a reaction sequence is carried out more than once
    d. A synthesis in which a reaction sequence is carried out only once
    e. A synthesis in which the first step is carried out twice

Answer: c Section: 6

SHORT ANSWER

29. Explain what is meant by a multistep synthesis. Give an example.

Answer:
A synthesis that requires more than one step to accomplish. For instance, adenine can be prepared starting with hydrogen cyanide and ammonia using a multistep synthesis.

Section: 2

30. Explain what is meant by linear synthesis? Give an example.

Answer:
Linear synthesis means building a molecule step-by-step from the starting material.
ex. $X \rightarrow Y \rightarrow Z \rightarrow P$

Section: 2

31. Design two different routes for synthesizing 1-aminobutane, $CH_3CH_2CH_2CH_2-NH_2$, using 1-butene, $CH_3CH_2CH=CH_2$, as the starting material.

    Answer:
    Route I

    $$CH_3CH_2CH=CH_2 \xrightarrow[\text{Peroxide}]{HBr} CH_3CH_2CH_2CH_2Br \xrightarrow{N_3^-} CH_3CH_2CH_2CH_2N\overset{\oplus}{=}N\overset{\ominus}{=}N$$

    $$\xrightarrow[Pt]{H_2} CH_3CH_2CH_2CH_2NH_2$$

    Route II

    $$CH_3CH_2CH=CH_2 \xrightarrow[\text{2. } HO^{\ominus},H_2O_2,H_2O]{\text{1. } BH_3} CH_3CH_2CH_2CH_2OH \xrightarrow[H_2SO_4]{K_2Cr_2O_2}$$

    $$\underset{CH_3CH_2CH_2COH}{\overset{O}{\overset{\|}{}}} \xrightarrow{SOCl_2} \underset{CH_3CH_2CH_2CCl}{\overset{O}{\overset{\|}{}}} \xrightarrow{NH_3}$$

    $$\underset{CH_3CH_2CH_2CNH_2}{\overset{O}{\overset{\|}{}}} \xrightarrow[\text{2. } H_2O]{\text{1. } LiAlH_4} CH_3CH_2CH_2CH_2NH_2$$

    Note: Other routes are also possible.
    Section: 1

32. What does an open arrow, $\Rightarrow$, represent in organic synthesis? Give an example.

    Answer:
    Open arrows represent a retrosynthesis operation.

    Ex. $A \Rightarrow B$

    This implies that A is the target molecule, and B is the starting material.
    Section: 2

33. What is the synthetic equivalent of $^+NO_2$?

    Answer: HNO_3/H_2SO_4 Section: 2

34. Perform retrosynthetic analysis to accomplish the following:

    Answer:

    oxidation

    Hydroboration
    oxidation

    Mg

    HBr   Peroxide

    Section: 2

35. Perform retrosynthetic analysis to accomplish the following:

    Answer:

    H^+ heat

    Mg

    LDA

    PBr$_3$

    Section: 2

36. Suggest a mechanism for the following reaction, the Darzens condensation:

Answer:

Section: 2

37. What is meant by an <u>extrusion</u> reaction? Give an example.

Answer:
A reaction in which a neutral molecule is eliminated. (e.g., CO_2, CO, N_2, SO_2, etc.)

Ex.

Section: 3

38. Two equivalents of Grignard reagent are required in the following reaction. Give an explanation.

Answer:
The first equivalent of RMgX is consumed by reacting with the labile proton of the OH group.

The second equivalent of RMgX will attack the carbonyl carbon and later be removed by hydrolysis.

Section: 1

39. Explain what is meant by using a <u>chiral</u> <u>auxiliary</u>?

Answer:
This is a method of synthesizing a single enantiomer by using an enantiomerically pure compound which, when attached to the reactant, causes the product with the desired stereochemistry to be formed.

Section: 5

40. Propose a mechanism for the following conversion:

Answer: This is a Diels-Alder (4+2) cycloaddition reaction.

Section: 6

41. Propose a mechanism for the Favorski ring contraction reaction shown below:

Answer:

Section: 6

42. Provide the reagents necessary to complete the following transformation.

Answer:
1. Br$_2$, hυ
2. H$_2$O, Δ
or
1. Br$_2$, hυ
2. NaOCH$_3$, CH$_3$OH
3. H$_3$O$^+$ or Hg(OAc)$_2$, H$_2$O; NaBH$_4$

Section: 2

43. Provide the reagents necessary to complete the following transformation.

Answer:
1. NaOCH$_3$, CH$_3$OH
2. MCPBA or CH$_3$CO$_3$H
3. H$_3$O$^+$ or $^-$OH

Section: 2

44. Provide the reagents necessary to convert 3-methyl-2-butanol to 2-methyl-2-butanol.

Answer:
1. conc. H$_2$SO$_4$ or PBr$_3$; NaOCH$_3$, CH$_3$OH
2. H$_3$O$^+$ or Hg(OAc)$_2$; H$_2$O NaBH$_4$

Section: 1

45. Provide the reagents necessary to accomplish the following transformation.

Answer:
1. O$_3$; CH$_3$SCH$_3$ or hot KMnO$_4$, $^-$OH
2. PhMgBr
3. H$_3$O$^+$

Section: 1

719

46. Provide the structure of the major organic product in the reaction shown below.

1. Mg, Et$_2$O
2. H$_2$CO
3. H$_3$O$^+$

Answer:

Section: 1

47. Describe a sequence of reactions by which E-5-bromo-2-pentene can be straightforwardly prepared from propyne.

Answer:
1. NaNH$_2$
2. ethylene oxide
3. Na, NH$_3$
4. PBr$_3$

Section: 1

48. Describe a sequence of reactions by which Z-3-hexen-1-ol can be straightforwardly prepared from 1-butyne.

Answer:
1. NaNH$_2$
2. ethylene oxide
3. H$_2$, Lindlar's catalyst

Section: 1

49. Describe a sequence of reactions by which the compound shown below can be straightforwardly prepared from acetylene.

Answer:
1. NaNH$_2$
2. CH$_3$CH$_2$Br
3. NaNH$_2$
4. cyclopentanone

Section: 1

50. Provide a series of synthetic steps by which the compound below can be prepared from benzene.

Answer:
1. $(CH_3)_2CHCl$, $AlCl_3$
2. HNO_3, H_2SO_4
3. Br_2, hυ

Section: 1

51. Provide a series of synthetic steps by which 4-phenyl-4-heptanol can be prepared from benzene.

Answer:
1. Br_2, $FeBr_3$
2. Mg, ether
3. $CH_3CH_2CH_2COCH_2CH_2CH_3$
4. H_3O^+

52. Propose a Wittig reaction-based synthesis of 3-hexene using propene as the carbon source and any other reagents necessary.

Answer:
Prepare 3-hexene from propene as follows:
1. BH_3THF
2. H_2O_2, NaOH
3. PCC
4. $CH_3CH_2CH=PPh_3$

Prepare the ylide $(CH_3CH_2CH=PPh_3)$ from propene as follows:

1. HBr, peroxides
2. PPh_3
3. BuLi

Section: 2

53. Provide the sequence of synthetic steps necessary to convert cyclohexanone into the compound shown below.

Answer:
1. $NaOCH_2CH_3$, $CO(OCH_2CH_3)_2$
2. $NaOCH_2CH_3$, $CH_2=CHCOCH_3$

Section: 1

54. Suggest a sequence of synthetic steps through which phenylacetic acid can be prepared from toluene and in which Grignard chemistry is employed.

Answer:
1. Br_2, hυ, <u>or</u> NBS
2. Mg, ether
3. CO_2
4. H^+, H_2O

Section: 1

55. Suggest a sequence of synthetic steps through which p-toluic acid can be prepared from toluene.

Answer:
1. Br_2, $FeBr_3$
2. Mg, ether
3. CO_2
4. H_3O^+

Section: 1

56. Suggest a sequence of synthetic steps through which 2-phenylethanol can be prepared from toluene. One of your intermediates must be acarboxylic acid.

Answer:
1. Br_2, hυ <u>or</u> NBS
2. NaCN, acetone
3. H^+, H_2O
4. $LiAlH_4$

or

1. Br_2, hυ <u>or</u> NBS
2. Mg, ether
3. CO_2
4. H^+, H_2O
5. $LiAlH_4$

Section: 1

57. Propose a reasonable synthetic route to prepare cyclohexylacetic acid from methylenecyclohexane.

Answer:
1. HBr, peroxides
2. Mg, ether
3. CO_2
4. H^+, H_2O

or

1. HBr peroxides
2. NaCN, acetone
3. H^+, H_2O

Section: 2

58. Propose a synthesis of 2-phenylethylamine from toluene and any other necessary reagents.

Answer:
1. Br_2, hυ <u>or</u> NBS
2. NaCN
3. $LiAlH_4$
4. H_2O

Section: 2

59. Using ethanol as your only source of carbon compounds and using any other necessary inorganic reagents, propose a synthesis of acetonitrile.

Answer:
Beginning with ethanol,
1. $Na_2Cr_2O_7$, H_2SO_4
2. $SOCl_2$
3. NH_3
4. P_2O_5
Section: 1

60. Provide a sequence of steps by which methylcyclohexane could be converted to racemic compound below.

Answer:
1. Br_2, hυ
2. NaOH
3. RCO_3H
4. $CH_3CH_2CH_2MgBr$
5. H_3O^+
Section: 5

MULTIPLE CHOICE

61. Which of the following best describes the approach known as combinatorial synthesis?

a. An approach that focuses on the single best retrosynthetic pathway for a set of target compounds.
b. An approach that designs multiple pathways to a single lead target compound.
c. A technique that uses the concept of mass production to prepare a large number of closely related structures.
d. A technique which combines high and low pressure conditions in an effort tom maximize yield.
e. None of the above.

Answer: c   Section: 98

62. Which of the following sequences represent methods by which pentanoic acid can be prepared?

    a. Treatment of pentanal with chromic acid.
    b. Ozonolysis of cyclopentene with subsequent oxidation work-up.
    c. Treatment of 2-pentanol with pyridinium chlorochromate.
    d. Both a and b.
    e. Both b and c.

    Answer: a   Section: 1

63. Which of the following is <u>not</u> an umpolung synthon?

    a. I
    b. II
    c. III
    d. IV
    e. None of the above

I.

II.

III.

IV.

    Answer: c   Section: 2

SHORT ANSWER

64. Provide a sequence of steps by which 3-pentanone can be converted to 2-ethyl-2-hydroxybutanal.

    Answer:
    1. Dithianyllithium
    2. H_2O
    3. $HgCl_2$, acetonitrile, H_2O
    Section: 2

MULTIPLE CHOICE

65. In which of the following reactions is a carbon-carbon single or double bond <u>not</u> formed?

    a. Attack of a Grignard reagent on a ketone.
    b. Wittig reaction between an aldehyde and a phosphorus ylide.
    c. Attack of an acetylide ion on an epoxide.
    d. Formation of a cyanohydrin.
    e. Clemmensen reduction of an aryl ketone.

    Answer: e   Section: 2

SHORT ANSWER

66. Using 1-pentene as the only carbon source, provide a synthesis of 5-nonanone.

    Answer:
    Preparation of pentanal: 1. BH_3
    　　　　　　　　　　　　 2. H_2O_2, HO^-
    　　　　　　　　　　　　 3. PCC

    Preparation of butyl magnesium bromide: 1. O_3
    　　　　　　　　　　　　　　　　　　　 2. Zn
    　　　　　　　　　　　　　　　　　　　 3. $NaBH_4$
    　　　　　　　　　　　　　　　　　　　 4. PBr_3
    　　　　　　　　　　　　　　　　　　　 5. Mg

    Preparation of 5-nonanone from pentanal: 1. $CH_3CH_2CH_2CH_2MgBr$
    　　　　　　　　　　　　　　　　　　　　 2. H_3O^+
    　　　　　　　　　　　　　　　　　　　　 3. PCC

    Section: 2

MULTIPLE CHOICE

67. An enamine is the synthetic equivalent of _____.

    a. an alkene
    b. an alkyne
    c. an acid chloride
    d. a ketone
    e. an amide

    Answer: d   Section: 3

68. Which of the following routes provides the best pathway to a 1,5-dioxygenated compound?

    a. a Diels-Alder reaction wherein the dienophile is CH_2=CHCHO
    b. a Michael reaction between an enolate and an α,β-unsaturated carbonyl
    c. a condensation reaction between an enamine and an α-bromoketone
    d. oxidative cleavage of a cyclohexene derivative
    e. None of the above

    Answer: b   Section: 3

SHORT ANSWER

69. What are the two most common reagents that can be used to protect the amino group?

    Answer:
    Acetyl chloride protects the amino group as the amide, and di-tert-butyl dicarbonate protects the amino group as the carbamate.

    Section: 4

Chapter 30:   The Organic Chemistry of Drugs: Discovery and Design

SHORT ANSWER

1. What is a drug?

   Answer:
   A compound which interacts with a biological molecule and thereby
   triggers a physiological effect is called a drug.
   Section: 98

MULTIPLE CHOICE

2. Which of the following terms is (are) synonymous with the term
   "proprietary name?"

   a. generic name
   b. trademark
   c. trade name
   d. brand name
   e. both c and d

   Answer: e   Section: 1

SHORT ANSWER

3. The _____ name of a drug is the name that any pharmaceutical company can
   use to identify the drug.

   Answer: generic   Section: 1

4. In pharmaceutical research, what is a "lead compound"?

   Answer:
   A lead compound is a pharmaceutically active compound of known structure
   which is used as the prototype in a search for other, structurally
   similar, active compounds.
   Section: 2

5. In pharmaceutical research, explain what is meant by "molecular
   modification."

   Answer:
   Molecular modification is the process through which analogs of the lead
   compound are synthesized by altering its structure (changing functional
   groups, ring sizes, chain lengths, etc).
   Section: 2

MULTIPLE CHOICE

6. What compound served as the lead compound in the research which developed today's most commonly utilized local anesthetics?

   a. morphine
   b. cocaine
   c. codeine
   d. sulfanilamide
   e. nitroglycerin

   Answer: b Section: 3

SHORT ANSWER

7. Novocain and Xylocaine are two compounds which are commonly used as local anesthetics. The structures of these two drugs are similar, but where Novocain has an aromatic ester functionality, Xylocaine has an aromatic amide group. Which of these drugs has a longer lifetime in the body and why?

   Answer:
   Xylocaine has the longer lifetime because amides are slower to hydrolyze than esters.
   Section: 3

MULTIPLE CHOICE

8. What compound is the most widely used analgesic for severe pain?

   a. Librium
   b. dextromethorphan
   c. Ativan
   d. Benadryl
   e. morphine

   Answer: e Section: 3

SHORT ANSWER

9. Based on the structures of morphine, codeine, and heroin shown below, explain why heroin is the most rapid acting member of the opiate family.

> Structures found together in Section 29.3

Answer:
Because heroin is less polar than morphine or codeine, it crosses the blood-brain barrier more rapidly and results in a more rapid physiological response.
Section: 3

10. What is the name of the receptor in the central nervous system which binds morphine and codeine?

Answer: the opiate receptor   Section: 3

MULTIPLE CHOICE

11. Who is credited with carrying out the first blind screen in his search for a "magic bullet" to cure African sleeping sickness?

   a. Jonas Salk
   b. Louis Pasteur
   c. Edward Jenner
   d. Paul Ehrlich
   e. Gerhard Domagk

Answer: d Section: 4

12. An important part of a random screen is the recognition of effectiveness in the compounds screened. What term is used to identify the test by which effectiveness is measured?

   a. assay
   b. dissolution
   c. modeling
   d. combinatorial analysis
   e. synergism

Answer: a Section: 4

13. What term is used to describe a screening test which is conducted outside the body, in a test tube for example?

a. *in toto*
b. *in tubo*
c. *in vitro*
d. *in loco*
e. *in vivo*

Answer: c Section: 4

SHORT ANSWER

14. What gave scientists the idea that azo dyes might be candidates as antibacterial agents?

Answer:
The fact that these dyes bound to wool fibers which are composed of protein led scientists to speculate that they might also bind to bacterial proteins.
Section: 4

15. Prontosil, a bright red dye, was an active antibacterial agent <u>in vivo</u> but not <u>in vitro</u>. Scientists also noted that mice given Prontosil did not excrete a red compound. What possible explanation of the activity of Prontosil is consistent with these two observations?

Answer:
The dye is converted into an active compound via some metabolic process of the mouse.
Section: 4

MULTIPLE CHOICE

16. What term describes a drug that, once taken, inhibits further growth of bacteria?

a. generic drug
b. bacteriostatic drug
c. antiviral agent
d. vasodiolation agent
e. bactericidal drug

Answer: b Section: 4

17. Which of the following drugs is a member of the benzodiazepine family?

    a. nitroglycerin
    b. acetylcholine
    c. morphine
    d. penicillin
    e. Librium

    Answer: e Section: 5

18. Which of the following drugs, discovered serendipitiously, is known to relieve angina pectoris by dilating cardiac blood vessels?

    a. nitroglycerin
    b. morphine
    c. acyclovir
    d. sulfanilamide
    e. Librium

    Answer: a Section: 5

SHORT ANSWER

19. What term is given to the specific cellular binding site with which some drugs interact to trigger their physiological effect?

    Answer: receptor Section: 6

20. What structural feature do most effective antihistamines have in common with histamine and how is this common feature believed to be related to the effectiveness of antihistamines as drugs?

    Answer:
    Like histamine, antihistamines have a protonated amino group at the end of an alkyl chain which allows these compounds to bind to a negatively charged portion of the histamine receptor. Antihistamines thus bind to the histamine receptor and prevent histamine from doing so.
    Section: 6

MULTIPLE CHOICE

21. What name is given to the receptors involved in neurotransmission, memory, and wakefulness?

    a. cholinergic receptors
    b. beta-blockers
    c. histamine receptors
    d. prostaglandin receptors
    e. opiate receptors

    Answer: a Section: 6

22. What type of enzyme is secreted by penicillin resistant bacteria which hydrolyzes penicillin before it has an opportunity to interfere with cell wall synthesis?

    a. esterase
    b. β-lactamase
    c. protease
    d. oxidase
    e. reductase

    Answer: b Section: 7

23. What new functional group is introduced to the penicillin structure when it is converted to a derivative by the oxidizing agent MMPP?

    a. N-oxide
    b. oxime
    c. sulfone
    d. amide
    e. carboxylic acid

    Answer: c Section: 7

24. What term is given to the effect which results when two drugs are used in combination and a response greater than the sum of the drugs' individual effects is observed?

    a. additive
    b. combinatorial
    c. distributive
    d. synergistic
    e. antagonistic

    Answer: d Section: 7

25. Name the last major class of antibiotics which were discovered. Members of this family work by inhibiting an enzyme required for transcription in bacteria.

     a. sulfa drugs
     b. cephalasporins
     c. tetracyclines
     d. fluoroquinolones
     e. penicillins

Answer: d   Section: 7

SHORT ANSWER

26. If we have a variety of effective antibiotics, why do pharmaceutical companies continue to search for new ones?

Answer:
Bacterial strains develop resistance to an antibiotic within 15 to 20 years. Drug resistant strains will only be fought with new antibiotics or synergistic agents.
Section: 7

27. How is the therapeutic index of a drug defined?

Answer:
The therapeutic index of a drug is the ratio of the lethal dose to the therapeutic dose.
Section: 8

28. Given a choice between two drugs of comparable effectiveness, would you choose the drug with the higher or lower therapeutic index? Explain.

Answer:
Choose the drug with the higher therapeutic index. The higher the therapeutic index, the greater the margin of safety associated with the drug's use.
Section: 8

29. In the context of drug design and synthesis, what is a suicide substrate?

Answer:
A suicide substrate is a drug which binds to the active site of a particular enzyme and subsequently reacts with the enzyme to inactivate it irreversibly.
Section: 8

30. The technique of relating a property of a series of compounds to biological activity is known as a QSAR. For what does this acronym stand?

    Answer: quantitative structure-activity relationship Section: 9

MULTIPLE CHOICE

31. What name is given to the strategy in pharmaceutical research which employs the concept of mass production, namely that a large group of related compounds can be prepared by covalently linking sets of building blocks?

    a. combinatorial synthesis
    b. serendipity
    c. random screening
    d. blind screening
    e. molecular modeling

    Answer: a Section: 11

32. Most antiviral drugs are analogs of _____ .

    a. carbohydrates
    b. steroids
    c. nucleosides
    d. proteins
    e. vitamins

    Answer: c Section: 12

SHORT ANSWER

33. What name is given to drugs which have limited demand and thus prove costly to develop?

    Answer: orphan drugs Section: 13

MULTIPLE CHOICE

34. Which of the following common drugs is an antihistamine?

    a. Claritin
    b. Norvasc
    c. Prozac
    d. Trimox
    e. None of the above

    Answer: a   Section: 98

35. Synthroid is used to treat underactivity of the _____.

    a. kidneys
    b. hypothalamus
    c. liver
    d. pancreas
    e. thyroid

Answer: e   Section: 98

SHORT ANSWER

36. Why does injecting a mixture of epinephrine and Xylocaine give a more prolonged anesthetic effect than does injecting Xylocaine alone?

Answer:
Because epinephrine is a vasoconstrictor, it reduces the blood supply to the injected area. This allows the Xylocaine to remain in the injected area for a longer period of time.
Section: 3

37. List 3 members of the opiate family, and give the main reason for their prescription by physicians.

Answer:
morphine, codeine, dextromethorphan, etorpine, pentazocine; their major use is as pain killers or analgesics
Section: 3

38. Provide the structure of histamine, and predict the most likely charge of the histamine receptor.

Answer:
Since histamine itself is positively charged, the histamine receptor is most likely negatively charged.

Section: 6

39. With what other receptor does the histamine receptor show overlapping activity? Give an example.

Answer:
The histamine receptor shows overlap with the cholinergic receptor. This is why the antihistamine diphenhydramine has been used to treat insomnia and to combat motion sickness.

Section: 6

MULTIPLE CHOICE

40. Which of the following drugs does <u>not</u> contain a heterocycle as part of its structure?

a. Benadryl
b. Lipitor
c. Prilosec
d. Morphine
e. Librium

Answer: a   Section: 6

41. Which of the following compounds was used as a lead compound in the development of antiulcer agents?

a. penicilloic acid
b. cocaine
c. heroin
d. 4-methylhistamine
e. β-lactamase

Answer: d   Section: 6

42. Which of the following reagents could be used to produce a sulfone from penicillin?

a. CH_3CO_3H
b. Zn, HCl
c. Raney Ni
d. H_3O^+
e. None of the above

Answer: a   Section: 7

SHORT ANSWER

43. What antibiotic was approved by the FDA in April 2000, to what family of organic compounds does it belong, and why was its approval particularly important?

    Answer:
    Zyvox, a member of the oxazolidinone family of compounds, was approved by the FDA in April 2000. Its approval was important because it was found to cure 75% of patients whose bacterial infections were resistant to all other antibiotics.

    Section: 7